WORLD® AIR POWER

J O U R N A L

Aerospace Publishing Ltd
AIRtime Publishing Inc.

Published quarterly by
Aerospace Publishing Ltd
179 Dalling Road
London W6 0ES
UK

ISSN 0959-7050

Aerospace ISBN 1 874023 64 6
(softback)
1 874023 65 4
(hardback)
Airtime ISBN   1-880588-07-2
(hardback)

Published under licence in USA and
Canada by AIRtime Publishing Inc.,
USA

Editorial Offices:
WORLD AIR POWER JOURNAL
Aerospace Publishing Ltd
3A Brackenbury Road
London W6 0BE   UK

Publisher: Stan Morse
Managing Editor: David Donald
Editor: Jon Lake
Associate Editor: Robert Hewson
Sub Editor: Karen Leverington
Editorial Assistant: Tim Senior

Origination and printing by
Imago Publishing Ltd
Printed in Singapore

Europe Correspondent:
Paul Jackson
Washington Correspondent:
Robert F. Dorr
USA West Coast Correspondent:
René J. Francillon
Asia Correspondent:
Pushpindar Singh
Canada Correspondent:
Jeff Rankin-Lowe

The editors of WORLD AIR
POWER JOURNAL welcome
photographs for possible publication,
but cannot accept any responsibility for
loss or damage to unsolicited material.

The publishers gratefully acknowledge
the assistance given by the following
people:

Joe W. Stout of Lockheed Martin
Tactical Aircraft Systems for his
assistance in compiling the *Lockheed
Martin F-16 Operators* and *Reconnaissance
Falcons* briefing.

Colonel Carl Herbertsson, of
Flygflottilj F5, and Mr Peter Liander for
their repeated assistance with the
Saab 105/Sk 60 Variant Briefing feature.

Gordon (K.G.) Hodson, for his
invaluable assistance with the Hawk
feature published in the last isssue of
World Air Power Journal.

**World Air Power Journal is
published quarterly and is
available by subscription and
from many fine book and hobby
stores.**

**SUBSCRIPTION AND BACK
NUMBERS:**

**UK and World (except USA and
Canada) write to:**
Aerospace Publishing Ltd
FREEPOST
PO Box 2822
London
W6 0BR
UK

**(No stamp required if posted in
the UK)**

USA and Canada, write to:
AIRtime Publishing Inc.
Subscription Dept
120 East Avenue
Norwalk
CT 06851, USA
(203) 838-7979
Toll-free order number in USA:
1 800 359-3003

**Prevailing subscription rates are
as follows:**
Softbound edition for 1 year:
$58.00
Softbound edition for 2 years:
$108.00
**Softbound back numbers
(subject to availability) are
$19.00 each. All rates are for
delivery within mainland USA,
Alaska and Hawaii. Canadian
and overseas prices available
upon request. American Express,
Discover Card, MasterCard and
Visa accepted. When ordering
please include your card
number, expiration date and
signature.**

Publisher, North America:
Mel Williams
Subscription Director:
Linda DeAngelis
Retail Sales Director:
Jill Brooks
**Charter Member Services
Manager:**
Janie Munroe

WORLD AIR POWER® JOURNAL

CONTENTS

Military Aviation Review

International

Three Eurofighters in the air

After being grounded for almost a year, BAe's EF 2000 development aircraft (DA.2, ZH588) made the Eurofighter's first public appearance at the Paris air show in June 1995, albeit in the static park and for the first few days, having done insufficient flying since resuming flight development on 17 May to work up a display routine. EF 2000 progress was reinforced by resumption of the flight programme of DASA's DA.1 in late May at Manching. Italy's DA.3 – the first EF 2000 with definitive Eurojet EJ200 turbofans, developing over 20,250 lb (90 kN) for take-off – made its first flight at Caselle on 4 June.

The EJ200-powered BAe-built DA.4, which is the first two-seat EF 2000, is scheduled to fly before the year's end from Warton. It will incorporate full mission avionics systems, apart from the ECR-90 radar, which is planned for installation in the DASA-built DA.5. Spanish two-seater DA.6 will not have a full avionics suite, and is likely to be the next to actually take to the air. All seven EF 2000 prototypes may fly this year.

Delays of up to six months have been reported with another software-led Eurofighter programme, the ECR-90 multi-mode pulse-Doppler radar, for which GEC-Marconi Avionics is prime contractor. The ECR-90 is due for installation in the German-built DA.5, but technical problems are being encountered with both the radar and the DBA-designed radome.

Flight development resumed on 17 May with a flight by DA.2 at Warton. This followed an extended layup of the first two prototypes while their digital flight-control system software was upgraded and ground tested to extend the envelope. Prior to the grounding, which also covered a more extensive avionics installation in DA.2, they had completed only 17 sorties and fewer hours, in the hands of three different pilots in each country.

With development responsibility for the EF 2000's FCS, GEC-Marconi Avionics had delivered the upgraded software in February to Daimler-Benz Aerospace, the original design and integration authority, for verification and rig-testing at Ottobrunn. This took a considerable time, although technical director Martin Friemere had already warned that software for the EF 2000's quadruply-redundant FCS was more challenging than similar systems in Gripen and Rafale. Apart from the EF 2000's greater instability, its redundancy management system was designed to cope with two major failures without FCS degradation. This in fact represents a double fail-safe capability, and the FCS must then be able to sustain a third major malfunction while keeping the aircraft operational.

Progress remained slower than planned, partly due to bad weather. The routine and uneventful 17 May sortie, actually the ninth flight for DA.2, was to have been followed by an intensive flight-test schedule to achieve clearance for a Paris debut. More important was the requirement to make up some of the lost time in its protracted develop-

The third Eurofighter to take to the air was Italy's DA.3, the first aircraft to be powered by the EJ200 engines. The aircraft first flew from Caselle on 4 June 1995.

ment programme, but DA.2's arrival at Le Bourget on 9 June was only its 18th flight in all, and its first cross-country.

Initially flown again for 1 hour and 25 minutes by BAe Military Division chief test pilot Chris Yeo, who was at the controls for its first flight on 6 April 1994, DA.2 arrived at Le Bourget in the hands of EF 2000 Project Pilot John Turner, accompanied by a Tornado chase-plane. Both aircraft returned to Warton on 12 June. Two days later, DA.2 extended its flight envelope into the supersonic regime.

The Paris air show was also to have been the venue for the signing of an MoU committing the Eurofighter team to the next EF 2000 R&D milestone. German parliamentary delays, however, in approving supplementary payments of DM531 million ($375.7 million) to Daimler-Benz Aerospace for certain Eurofighter cost overruns resulted in a similar deferral of the MoU signature. The UK is seeking an increase in its EF 2000 production share to at least 42.5 per cent from its originally planned 33 per cent, mainly at the expense of Germany, since it is the only one of the Eurofighter partners to adhere to its initial commitment to 250 aircraft. Germany was also to have purchased 250 EF 2000s, in return for a similar 33 per cent investment, but its funding has now been restricted to a maximum of 140.

JAS 39 Gripen agreement between BAe and SAAB

The appearance of the Eurofighter and a Gripen (39-4) at Paris was in support of the PR push surrounding a joint announcement by Saab and BAe. The finalisation of plans by British Aerospace and SAAB for the joint development, production and marketing of an export version of the JAS 39 Gripen was announced at Le Bourget. BAe officially regards the Gripen as a light combat aircraft which could fill a perceived marketing gap between the Hawk 200 and the Eurofighter, and hopes to sell 200-400 Gripens in competition with fighters like the F-16 and MiG-29 over the next 15 years.

Work on export Gripens will be shared 55/45 per cent in Saab's favour, although the nature of the workshare and the assembly location have still to be agreed. British work on the Gripen will be conducted at BAe's Brough plant. Deliveries of the export Gripen are planned for 1999, subject to satisfactory integration with production plans for the first 140 for the Flygvapnet, including 14 two-seat JAS 39B combat trainers.

Western Europe

FRANCE:

First Hawkeye orders

Following the conclusion of 100 per cent offset arrangements (excluding powerplant costs) in November 1995, Aéronavale orders have been placed for two E-2C Hawkeyes. These were placed through an FMS contract. The aircraft will operate from the nuclear-powered carrier *Charles de Gaulle*. This will commission in 1999, and will operate Rafale Ms instead of F-8E(FN) Crusaders and Super Etendards.

Two more E-2Cs would be required for a second vessel, with an estimated package cost for the four E-2Cs of Fr7 billion ($1.43 billion). Total cost of the *Charles de Gaulle* and its aircraft is quoted as Fr71 billion ($14.5 billion).

GERMANY:

Tu-154 inspection roles

One of two Tupolev Tu-154Ms (11+02) taken over by the Luftwaffe from the former East German air force (NVK) has been modified in Dresden

The first of 64 Hornets for the Finnish air force flew for the first time at Lambert Field, St Louis, on 21 April 1995. Here it is seen on its second flight, undertaken on 23 May. The two-seater wears a small roundel and serial number on the front fuselage.

by Elbe Flugzeugwerke for surveillance and inspection flights. Up to 12 of these flights will be yearly, over the former Soviet Union, in accordance with the 1990 Open Skies treaty.

Three Zeiss Jena LMK 2015 optical and three Zeiss Oberkochen VOS-60 video cameras have been installed in the underfloor baggage compartment of the Tu-154 for the initial DM45 million ($32 million) conversion. These will be supplemented in 1997 by an additional panoramic camera, an infra-red linescanner and a Rossar-G side-looking synthetic-aperture radar for more comprehensive surveillance over Eastern Europe, Belarus, Ukraine and Russia as far east as the Urals, to a maximum permissible distance of some 3,500 nm (4,025 miles; 6477 km). Access will be permitted to four designated Russian airfields .

EGRETT prototype for sale

The prototype Garrett TPE331-powered Grob/E-Systems D500 EGRETT high-altitude surveillance aircraft is being offered for sale by Germany's VEBEG federal trust company, which handles military equipment disposals, after only 437 hours flying. The D500 was funded by the Federal Defence Ministry as the first of 14 aircraft planned to equip a Luftwaffe arms-verification overflight squadron.

At least three EGRETT 2 examples were built before the programme was cancelled in 1992, but on 31 March 1995 Grob flew the twin-turboprop Strato 2C derivative with a 185-ft (56.4-m) span and a planned 80-hour endurance. After flight development, this is scheduled for delivery to the German government for high-altitude surveillance and research in 1996.

ITALY:

Air force plans

Air and ground crews of the Italian air force (AMI) are training with No. 56 (Reserve) Squadron, RAF – the Tornado F.Mk 3 OCU – in readiness for the transfer of 24 Tornados on long-term lease from the RAF. These will serve pending deliveries of the first of 110-130 EF 2000s early in the next century. The first AMI unit to operate Tornado ADVs will be 12° Gruppo of 36° Stormo at Gioa del Colle, which already operates a maritime strike Tornado IDS squadron.

This will be followed by a second squadron of 12 Tornado ADVs – probably 21° Gruppo, 53° Stormo – at Trapani/Birgi in 1997. The AMI's Tornados will reinforce the two air defence wings of 90 F-104ASA Starfighters being upgraded with multi-role capabilities, plus an OCU with 18 similarly upgraded TF-104Gs. The AMI has outstanding requirements for four AEW aircraft and four tankers when funding permits.

NETHERLANDS:

Apache confirmed

French attempts to persuade the Dutch government to give preference to the European Eurocopter Tiger, to meet its long-standing attack helicopter requirement, were unsuccessful, following cabinet selection on 7 April of the MDH AH-64D Apache. An FMS order for 28 AH-64Ds, plus options on three more (all initially without Longbow mast-mounted fire-control radar) was given parliamentary approval in May, for delivery from early 1998. The US package also includes the loan of 12 ex-US Army AH-64As from early 1996 for crew and maintenance conversion training.

Official statements from the Hague said that selection of the Apache was based on crew safety, robustness, self-defence capability, navigation and night vision systems, operational flexibility and development risk. It was also claimed that the Tiger failed to meet the Dutch 11th Air Mobile Brigade's requirement for a minimum 2.5-hour mission endurance, although this was denied by Eurocopter. Even more critically, a Tiger production commitment – forecast for the end of June for the originally-planned 427 aircraft – had still to be finalised by France and Germany.

The DFl 1.3 billion ($807 million) contract was signed on 24 May – a unit cost of some $27 million. The Dutch AH-64Ds may have six or more mast-

mounted Martin Marietta Longbow millimetre-wave fire control radar systems delivered at a later date. McDonnell Douglas Helicopter Systems are placing initial offset contracts worth DFl895 ($570) million from a requirement of 100 per cent of the contract value, although Eurocopter claimed its Tiger offset proposals would have been for 120 per cent of the package cost.

An official Dutch report claimed that 34 Tigers, each carrying only 12 ATMs, would have to be bought to do the work of 30 AH-64Ds. The AH-64D weighs nearly twice the Tiger's weight of 5547 kg (12,230 lb), and also has marginally cheaper 20-year lifecycle costs of DFl3.16 ($1.96) billion. In later testimony to the Dutch parliament, before endorsement of the AH-64D, Eurocopter co-chairman Jean-Francois Bigay countered that Tiger 30 year lifecycle costs would be DFl1 billion ($620 million) less than Apache's.

F-16 and P-3 upgrades

Pratt & Whitney has received a $90 million contract for upgrade kits to modify the F100 turbofans of RNAF (KLu) F-16 fighters to -220E power and durability standards. Earlier this year, P&W Government Engines and Space Propulsions Customer Support & Services Department began supplying 106 engine kits for the local upgrade of 92 of the RNAF's F-100-PW-200-powered F-16A/Bs. This will bring these aircraft to a similar standard to the 42 RNAF F-16s already fitted with F100-220 engines. Apart from increasing core life and inspection intervals, the F100-220E features digital engine controls and fault isolation capabilities.

The Dutch government has allocated DFl209 million ($130 million) for a Capability Upkeep Programme (CUP) to upgrade 13 Lockheed P-3 Update IIs

Above: An important weapon for the Rafale is the MATRA Apache stand-off sub-munitions dispenser, seen here during captive-carry tests from the Rafale B 01 prototype.

Right: A new type for the French army (ALAT) is the SOCATA TBM 700, used for liaison duties. The aircraft, originally developed in conjunction with Mooney, is also used by the Armée de l'Air.

of the Royal Netherlands navy (MLD). To be completed between 1997 and 2002, the CUP will upgrade the radar, ESM, acoustic sensor, data processing and communications systems, as well as adding forward-looking infra-red and ultra-violet sensors

PORTUGAL:

Alpha Jet units formed

Of the 50 Alpha Jets transferred from the Luftwaffe to the FAP in late 1993, five are being stripped for spares, and the remaining 45 are being operated by Esquadras 103 and 301 from Beja air base (BA 11). In Esq 103, 15-20 Alpha Jets have replaced the 12 Northrop T-38s and older T-33s previously used for advanced training (retired in September 1991 and June 1993, respectively), while Esq 301 is operating up to 25 Alpha Jets in the weapons training/ground attack roles undertaken by Fiat G.91R/3s. The Alpha Jets retain the advanced nav/attack systems, while those of Esq 301 are also fitted with the belly-mounted 27-mm Mauser cannon pod.

FTB.337 disposals

The Portuguese air force (FAP) has disposed of 11 Reims-Cessna FTB.337G forward air control aircraft from 32 originally delivered, most of which were withdrawn from use some years ago. They have been stored at the former Ota airfield (BA-2), near Lisbon, and have been sold to an American dealer. One or two FTB.337s were operated for liaison and light transport roles until the end of 1995.

TURKEY:

IAI receives Phantom 2000 upgrade contract

After delays resulting from Turkish industry's insistence on receiving a greater workshare, Israel Aircraft Industries is being awarded a $600 million contract to upgrade 59 McDonnell Douglas F-4Es of the THK to Phantom 2000 standards, after strong competition from Daimler Benz Aerospace which offered an upgrade based on the Luftwaffe F-4F ICE modernisation. Originally developed for Israel's own F-4 force, the Phantom 2000 upgrade involves the installation of a new Westinghouse Norden Systems APG-76 multi-mode radar, Kaiser/El-Op wide-angle HUD, and other improved digital avionics. Other changes include structural reinforcements to improve fatigue life, and a complete rewiring. Turkish Phantom 2000s will have small strakes added above the intake flanks to improve combat manoeuvrability.

UNITED KINGDOM:

WE 177 withdrawal

A Parliamentary statement earlier this year confirmed the scheduled withdrawal from service by late 1998 of the RAF's WE 177 freefall nuclear weapons. These have been the primary weapon for the Tornado IDS, and for nuclear-assigned Vulcans, Jaguars and Buccaneers before that. Estimated at between 180 and 200 in number, the WE 177 designation covered a family of weapons using the same shell and physics package, with differing amounts

Military Aviation Review

Above: The Armée de l'Air operates a pair of A310-304s on transport duties with ET 3/60. They were previously flown by Royal Jordanian Airlines, and may yet be converted for tanker duties.

Below: The first six AH-64As destined for Greece (EΣ1001-1006) passed through Lakefront Airport, in early January 1995, on their way to New Orleans port for onward shipment. Seen here is EΣ1002.

of tritium to vary the yield. The UK's stockpile of 155-175 tactical 600-lb (431-kg) WE 177A/Bs will be replaced by single- and multiple-warhead versions of RN Trident D-5 SLBMs in 1998, to maintain the UK's future nuclear deterrent. The WE 177 was developed from US B57 and B61 nuclear weapons. The A and B variants respectively used variable-yield warheads of up to 200- and 400-kT yield, for tactical or sub-strategic use. The stockpile of WE 177s was reduced by 50 per cent with the withdrawal of the Laarbruch Tornado Wing, after an announcement by the Minister of Defence on 14 October 1991.

Until its 1992 withdrawal, the 950-lb (430-kg) WE 177C, with a 4- to 20-kT yield, was available as a depth bomb or as a freefall surface attack weapon. It was intended mainly for use against maritime and submarine targets by RN Sea King helicopters and Sea Harriers, as well as RAF Buccaneers and Nimrod ASW patrol aircraft.

The RN's Trident submarines are to be armed with up to 200 1,000-nm (1,115-mile; 1850-km) extended-range GPS-equipped Block 3 Tomahawk cruise missiles with 750-lb (340-kg) conventional HE warheads. A formal request for these, costing around $200 million, is to be made to the US government later in 1995, following completion of feasibility studies.

AEW Sea King upgrade

After three years in the pipeline, tenders were invited in late May for a mission systems update (MSU) for the FAA's 10 Sea King AEW.Mk 2A Airborne early-warning helicopters. Main element of the MSU will be a new pulse-Doppler radar, to replace the existing Thorn-EMI Searchwater system.

The upgraded Searchwater 2000 is one of several new submissions which also include GEC-Marconi Radar Systems division's Blue Vixen radar (already equipping RN Harrier F/A.2s), and radars from Lockheed Aircraft Service Co. and Thomson-CSF. Other upgrades planned for the projected Sea King AEW.Mk 7 include a new central tactical system, an integrated Mk XII IFF, and a JTIDS terminal. The programme will also involve three planned AEW.Mk 5 conversions (from HAS.Mk 5 airframes), to cover anticipated attrition.

Jet Provost engines recycled for Jindivik target drones

Twenty surplus Viper 202 turbojets from retired Jet Provosts are being overhauled and converted to Mk 201C standard by Rolls-Royce under a £2.2 billion MoD contract. Rated at 2,500 lb (11.12 kN) of thrust for take-off, they are to be installed in 18 new and

upgraded Jindivik Mk 4A target drones. These were ordered from the Systems Division of Aerospace Technologies of Australia (AsTA) through a $A36 million ($26 million) contract in mid-1994, the last of 516 Jindiviks built since 1951.

With a speed of 540 kt (620 mph; 1000 km/h) and a ceiling of 65,800 ft 20056 m), 249 Jindiviks have been expended since 1960 as targets at the Aberporth weapons range. Ten are currently operated from the adjacent airfield of Llanbedr, and eight remain to be assembled from earlier deliveries.

In UK service, the Jindiviks normally tow twin dart targets on 300 ft (91 m) of cable, using radar or infra-red augmentation when required to simulate various aircraft. They are fitted with missile proximity camera recorders in wingtip pods. Towed targets ensure a high survival rate, one Jindivik having completed no fewer than 324 sorties.

Development and component manufacture for the Viper 202 conversions is being undertaken by Rolls-Royce's Bristol facility, with final assembly and initial testing by RR Aero Engine Services at Ansty. Delivery to the MoD of the 18 new Jindiviks from AsTA's re-opened drone production line, upgraded in the UK with digital radio command and telemetry systems, is due for completion in 1996-97, when they will double the current Mk 4A UAV strength at Llanbedr.

Jindiviks have been extensively used for weapons trials by the Australian armed forces. Ten were exported to Sweden in 1959-60, while 42 Jindivik Mk 3As were operated for the USN by Ling-Temco-Vought from late 1966.

Harrier T.Mk 10s in service

The RAF's first (of a planned 13) Harrier T.Mk 10s are operational with No. 20 (R) Sqn, the Wittering-based Harrier OCU. ZH657 arrived on 30 January from Boscombe Down. All three OCU T.Mk 10s have the nose-mounted Hughes Angle Rate Bombing System and GEC-Marconi Sensors FLIR of the GR.Mk 7, but retain the early 75 per cent LERX. They had been delivered by mid-February, and were followed by another for No. 1 Sqn. Three more are going to Nos 3 and 4 Squadrons in Germany, and another two will initially go to Boscombe Down.

France has recently purchased the Paveway II series of laser-guided bombs for its tactical air forces. Here a 500-lb class GBU-12 is released by a CEAM 330 Mirage F1CT.

Flight amalgamation

1 April marked the inauguration of No. 32 (The Royal) Squadron, RAF, at Northolt, combining the aircraft of the former Queen's Flight from Benson with No. 32 Squadron's original VIP and government communications aircraft. In its revised form, No. 32 (The Royal) Squadron, commanded by Wing Commander Andy Barratt, operates three ex-Queen's Flight BAe 146 CC.Mk 2s (BAe 146-300) in A Flight; two ex-Queen's Flight Westland Wessex HCC.Mk 4s and four Westland/Aérospatiale Gazelle HT.Mk 3s in B Flight; and two BAe 125 CC.Mk 2s (BAe 125-600) and six CC.Mk 3 (BAe 125-700) in C Flight. Servicing and maintenance is undertaken under contract by the Flight Refuelling Aviation subsidiary of Cobham PLC. This is also undertaking the respray of four BAe 125s from their current tactical low-visibility satin grey to the traditional Queen's Flight scheme of high-gloss white, with red and blue trim.

Test crew escape EH101

Parachute escapes from helicopters are rare, but four crew members baled out from Westland's EH101 Merlin PP4 prototype during a test flight from Yeovil on 7 April. Only minor injuries were incurred by test pilot John Dickens, who stayed with the aircraft until about 1,200 ft (366 m), after ordering his co-pilot, Don Maclaine, and two test observers to bale out at 12,000 ft (3658 m). The aircraft crashed in a field in a level attitude, and maintained much of its structural integrity, with no post-impact fire. Reports that the EH101 was rotating laterally as it descended suggested a tail rotor problem.

As one of nine pre-production EH101s, PP4 (ZF644) was the first to be fitted with the Rolls-Royce/Turboméca RTM322 engines specified for the 44 Merlins and 22 similar Griffon HC.Mk 1 transport versions ordered for the RN and RAF, respectively. ZF644 had been flying since June 1989, although only since July 1993 with its new powerplants. Some 3,600 hours of flight-testing had been accumulated by the EH101 fleet up to the time of the accident, with only another 150 hours or so planned to complete the initial development programme. The RN is due to receive its first Merlins after Boscombe Down clearance starts in late 1996, but funding is still awaited by the Italian navy for its procurement of 16 ASW-equipped and GE T700-T6A-powered EH101s.

This was the second loss to the EH101 test fleet, PP2 having crashed in Italy on 21 January 1993 with the loss of four lives, following uncommanded application of the upgraded naval rotor brake in flight.

Skyvans replace balloons

Two civil Short Skyvan 3Ms are being operated by Hunting Aviation from Kidlington Airport, on an initial three-year contract for service paratroop training. This was previously undertaken from tethered kite balloons. With a rear ramp and a capacity for up to 14 paratroops, the Skyvans are well suited to parachute training, and Kidlington is only a few minutes flying from the RAF's main training drop-zone at RAF Weston-on-the-Green.

RAF Nimrod R.Mk 1P lost

One of the three Nimrod R.Mk 1Ps of No. 51 Squadron ditched in the Moray Firth during a test flight on 16 May. The Nimrod was being test flown from RAF Kinloss following a major inspection, when a fire started in its starboard engine bay.

Actuation of the built-in extinguishing system proved unsuccessful, and the captain, Flight Lieutenant Stacey, ditched 4 miles (6.5 km) short of RAF Lossiemouth before the wing structure burned through. All seven crew members were rescued, suffering only minor injuries. They were picked up by two Sea King HAR.Mk 3s of No. 202 Sqn exercising nearby.

The Nimrod R.Mk 1s are being upgraded through a 1992 fixed-price contract by Raytheon's E-Systems subsidiary with the installation of new digital and computerised electronic reconnaissance equipment codenamed Star Window. Main elements of the new equipment included two high-speed search receivers, a wide-band digital direction-finding system, 22 pooled digital intercept receivers, and colour active-matrix workstations. Most of the aircraft's mission equipment had been removed prior to its major inspection, and had not been refitted at the time of the accident. Such is the sensitivity of the Nimrod R.Mk 1's equipment that it is routinely removed before the aircraft leave their Waddington base for any form of servicing or engineering attention that cannot be provided at squadron level.

While 20 of the 25 remaining Nimrod MR.Mk 2s equipping the RAF's three maritime patrol squadrons are now all based at Kinloss, the R.Mk 1s

operate from Waddington, alongside the seven Boeing E-3Ds of No. 8 Sqn, having moved there on 28 April with the closure of RAF Wyton. Salvage of the Nimrod R.Mk 1P from the 70 ft (21.3 m) of water in which it sank was begun almost immediately.

The MoD's Staff Requirement (Air) 420 programme to replace the Nimrod early in the next century was launched in January 1995, with refurbishment and upgrading of the existing aircraft as one of several possible options.

Special mission Chinooks

Eight of the 14 Boeing Chinooks ordered in March will be HC.Mk 3 versions, configured for special forces use, and similar to the MH-47E. These are fitted with uprated Lycoming T55-L-714 turboshaft engines, AN/AAQ-16 FLIR and other equipment, including double-capacity fuel tanks and a refuelling probe.

RAF C-130J order details

As the prototype Hercules II, the first RAF C-130J-30 (c/n 5408) is now undergoing final assembly, together with two others for the RAF and the first two short-fuselage C-130Js for the USAF. Announced on 16 December 1994, the RAF order is for 25 C-130J Hercules IIs, plus five options, and will initially include 15 extended-fuselage Dash 30 versions. The rest were ordered as short-fuselage C-130Js, but with an option to change these to -30s.

Roll-out of ship number 5408 is scheduled for 22 September, followed by a first flight in December. The first five Hercules IIs will then start flight development and certification clearance before deliveries from late 1996. C-130H production will end, to be replaced by the Hercules II at an initial rate of about three per month. Apart from increases in performance, with a maximum speed of 360 kt (667 km/h), the C-130J is designed for savings in life cycle costs. Ordered as a gap-filling aircraft pending a final decision on the FLA, the RAF's requirement for the C-130J relies on the 'fact' that the existing Hercules could not be refurbished for the interim period. It is thus surprising that some 25 RAF Hercules are being offered to Lockheed (in part exchange), who presumably feel that they do still have a viable life.

Eastern Europe

CZECH REPUBLIC:

Budget limits re-equipment

Budget limitations and initial allocations for the CKr1.7 billion ($65 million) procurement programme for 72 ITEC F124-powered Aero Vodochody L-159 single-seat light combat aircraft have resulted in the Czech air force having to scale down re-equipment plans. The government has vetoed air force proposals to acquire Western fighters on cost grounds.

The Czech air force will now be restricted to upgrading 24 of its MiG-21bis fleet, although allocations for this have also been reduced. Upgrades will be confined to NATO-standard communications and GPS, plus replacement of the Soviet SRZO-2 IFF by a Western unit.

POLAND:

New ground attack project

Western industry is involved with the PZL aerospace group in the joint development of the bizarre-looking PZL-230 Skorpion ground-attack fighter project, designed to replace Poland's ageing Sukhoi Su-22s early in the next century. Exhibited in model form for the first time in the West at the 1995 Paris air show, the radical PZL-230 follows the

example of the Fairchild A-10, mounting its two Lycoming LF507 turbofans between canted stabilators to shield the jet pipes.

Apart from Lycoming's contribution, AlliedSignal is supplying Bendix/King avionics for the Skorpion, while Lear Astronics is responsible for the fly-by-wire control system.

Design features on the PZL-230 model include a canard delta layout, with leading-edge slots and slotted flaps for short-field performance. Maximum take-off weight of some 10000 kg (22,046 lb) includes up to 4000 kg (8,818 lb) of external stores on 10 or more weapons pylons. Detailed design submissions are now being evaluated and a mock-up has been built.

ROMANIA:

Elbit leads MiG-21 upgrade

Elbit has now been revealed as prime contractor for the upgrade of about 100 Romanian air force MiG-21s, in conjunction with Aerostar as local subcontractor. Elbit-supplied components include multi-function monochrome and colour cockpit displays, integrated into a revised control panel; a raster head-up display; a computerised air data unit; a multi-role computer; HOTAS; and a DASH helmet-mounted sight. Romania's MiG-21 upgrade is

Above: Powered by the LHTEC T800 engines, the Agusta A 129 International also features a five-bladed rotor. The aircraft is shown carrying a 19-round rocket pod, gun pod and Stinger missile launchers.

Left: The Swedish army is in the process of replacing its elderly Hkp 3C (AB 204B) helicopters with the Hkp 11 (AB 412SP).

Above: Two BN2T Islander CC.Mk 2s are based at RAF Northolt for unspecified utility duties. The aircraft are operated by the base's Station Flight.

Below: First seen on Tornados of the Lossiemouth maritime wing, the mid-grey scheme is being worn by the Marham reconnaissance wing. This aircraft is from No. 13 Sqn.

believed to include an IAI Elta EL/M-2032 advanced fire-control radar.

Elbit's retrofit package has also been flown in an F-5, and is available for other aircraft, including the F-4. The F-5 upgrade is applicable to the T-38 trainer, and forms the basis for Elbit's submission for the modernisation of 425 USAF T-38s.

Bell AH-1F acquisition

Some 96 locally-produced Bell Textron AH-1F Cobras will be the first Western combat aircraft acquired by a former Warsaw Pact country. A letter of intent for these was signed by the Romanian government during the Paris air show.

Bell is planning to transfer to Romania some airframe production tooling and such items as transmissions and rotor blades might also be built in Romania, according to a Bell Textron spokesman, and its AH-1F acquisition would represent a major step towards that country's full NATO integration. Romania is expected to start replacing its Alouette III and Puma attack helicopters with AH-1Fs in 1999, and to receive its 96th Cobra in 2005. Some 470 AH-1Fs still serve in active, reserve and Guard units of the US Army.

RUSSIA:

New 'Flanker' debuts

The Sukhoi Design Bureau sent a relatively junior delegation to Paris, and, in the absence of charismatic Designer General Mikhail Simonov, left unanswered many questions about

the aircraft which they accompanied. Appearing at Paris as the Su-32FN, 'Bort 45' appeared identical to that hitherto known in the West as the T-10V or Su-34. Prior to the 1995 Paris air show, Sukhoi's Su-32FN designation had applied to a militarised, tricycle undercarriage-equipped version of the Su-29LL tandem-seat trainer, but for unexplained reasons this designation was then transferred to the third Su-34. Members of the delegation hinted that the designation change was recent and the result of an edict from Simonov himself. The Su-34 had formerly been described as a successor to the Su-24 strike aircraft, while the Su-32FN was presented by the Sukhoi OKB as being optimised for naval roles.

The side-by-side 'Flanker' was almost certainly developed to meet a Soviet requirement for an Su-24 'Fencer' replacement. The programme began with the conversion of a standard Su-27 as the Su-27IB technology demonstrator. This aircraft 'Bort 42' featured a standard 'Flanker' airframe, to which a two-seat side-by-side cabin and new forward fuselage had been grafted. It was briefly known as the Su-27KU in the West, indicating a carrier-based training application. Although photographed making a deck landing approach to the carrier *Admiral Kuznetsov*, the aircraft was never equipped for shipborne operation and the Su-27KU designation proved to be ingenious disinformation.

With a 4000-km (2,160-nm) range on internal fuel, the real forte of the side-by-side Su-27/32/34 is as a sub-strategic bomber, which is how the first purpose-built example 'Bort 43' emerged from Novosibirsk/Chkalov's

factory. That aircraft made its initial flight on 18 December 1993. This revealed a number of differences from earlier examples, including leading-edge root extensions aft of the canards as a continuation of the chined radar nose; broader-chord and thicker fins providing additional fuel stowage; a larger-diameter and considerably extended rear fuselage 'sting' capped by a radome; a reinforced nosewheel unit; and tandem-wheeled main landing gear bogies to cope with the increase in maximum take-off weight to no less than 44360 kg (97,800 lb).

Speculation was that the Su-34's main radar would be a development of Phazatron's Zhuk 27 coherent pulse-Doppler X-band radar, already flying in development Su-35s. The radar has a 5-kW peak power output, an average power of 1 kW, and a claimed detection range in air-to-air modes of 54 nm (100 km) in the forward hemisphere and 30 nm (55 km) aft. Air-to-surface modes of the Zhuk 27 include real-beam ground-mapping with Doppler sharpening; synthetic-aperture; MTI; automatic low-level terrain-following that is allegedly accurate to within 1 m (3.3 ft); ranging and aircraft velocity inputs for navigation system updates; and track-while-scan of up to four ground targets. The new Leninets St Petersburg V004 advanced radar has now been mentioned in connection with the Su-32FN, and may be competing with developments of the Su-35's Zhuk 27 or phased-array PH, or the rival NIIP Zhukhovskii radar which may be designated NO11 (although this designation has also been used in connection with the Zhuk 27).

The extended, large-diameter tail fairing can obviously house a bigger rear-facing radar (for cueing rearward-firing AAMs) than the Su-35's Phazatron NO14, with its limited range. The Su-34's rear-mounted radar is also thought to comprise an active part of its integrated warning system, with a range of up to 20 km (11 nm). Other elements include wingtip Sorbtsiya-S jamming pods – replaced by missile rails on '45' at Paris – and a MAK-UL missile-launch detection dome behind the cabin. Russian sources suggested that the tailcone of the Su-32FN actually housed a magnetic anomaly detector, but this seems most unlikely.

The third pre-series Su-34 shown at Le Bourget as the Su-32FN joined the flight development programme at Zhukhovskii about a year after the first, and will be followed by up to 10 others. Another has also been completed for ground and fatigue testing. No. 45 arrived in Paris non-stop from Moscow in 3.5 hours, without using its retractable air-refuelling probe.

Entered via a folding ladder in the nosewheel bay, and furnished with twin Zvezda K-36 ejection-seats, the cockpit is a 17-mm (0.67-in) thick titanium-armoured capsule said by Sukhoi General Designer Mikhail Simonov to be "as big as the Tu-134's". Big enough, in fact, to incorporate a toilet compartment and a galley, both essential accessories given the aircraft's extreme range.

Listed (mainly maritime) weapons for the Su-32FN included some previously unmentioned types, and comprised medium-range (up to 250 km/135 nm) Raduga Kh-41 ASM-MSS Moskit ramjet-cruising supersonic active radar anti-ship missiles, TV-command seeker-equipped AGM-TVC AShMs and Kh-31P passive anti-radar missiles, plus shorter-range infra-red and laser homing S-25IRS/LD and Kh-29L/T (AS-14 'Kedge') laser and TV-guided AGMs, and R-73E (AA-11 'Archer') IR-homing AAMs. All weapons are carried on a single large underfuselage beam or on underwing pylons. The Su-32FN retains the 30-mm GSh-301 of the Su-27s and -35s in the starboard wingroot, with 150 rounds to port.

Su-32/34 powerplants have not so far been confirmed but may be developed versions of the Saturn AL-35Fs in the Su-35. Maximum low-level speed of the Su-32 is quoted as a near-sonic 1400 km/h (755 kt), increasing to Mach 1.8 (2500 km/h; 1,350 kt) at altitude.

Lockheed/Yakovlev combat aircraft collaboration

A recent agreement between Lockheed Martin and the Yakovlev OKB in Moscow to co-operate on STOVL technologies relating to the US Joint Advanced Strike Aircraft (JAST) programme marks the first ever co-operation between America and Russia on a combat aircraft. Lockheed Martin is interested in Yakovlev's extensive experience gained over many years with the Yak-36, Yak-38 and supersonic Yak-41 V/STOL fighter programmes.

Although cancelled in 1991 after two prototypes had flown, the Yak-41 embodied advanced technologies in integrated fly-by-wire, autostabilisation and thrust-vectoring flight-control systems, plus afterburning in a swivelling engine nozzle. Yakovlev is supplying Lockheed Martin with data and information applicable to the US company's designs in preparation for the demonstrator phase of the JAST programme.

MiG-MAPO developments

The formation of a unified MiG-MAPO group by the amalgamation of the formerly discrete Mikoyan Design Bureau and the Moscow Aircraft Production Organisation factory (formerly known as Plant 30, or the Znamaya Truda production facility) received formal Russian government approval in May.

Completion of the development of the MiG-29M has now been funded by the Russian government. The aircraft has been redesignated MiG-33 (a designation already adopted for the export version of the MiG-29M), since it is claimed to be virtually a new aircraft. Development will be completed this year. After major integration problems the aircraft is "100 per cent ready" and remaining development will be restricted to confirming existing test results, which reportedly show that it meets its specification. The Bureau claims that it has received verbal assurances of a small production order. With a Phazatron Zhuk pulse-Doppler radar, more fuel, new weapons systems, fly-by-wire controls and many other improvements, the MiG-29M is claimed to be three to four times more effective than previous versions.

Details of 1-42 emerge

MiG-MAPO claims to have been working on stealth aircraft programmes for the past 10 years, and have undertaken computer studies and flight simulation of every current combat aircraft. These are being applied to the development of new combat aircraft for the next century, and some of these inputs have been incorporated in the next-generation Project 1-42. Some sources suggest that the designation 1-44 may now be appropriate for the prototype in its present configuration.

This designation still appears to be classified so far as Russia is concerned, although MiG-MAPO has admitted that work is progressing on a new fighter to match the US ATF (F-22) programme. The project began in 1983, initially in competition with a Sukhoi design. Drawings for the prototype were reportedly finalised in 1987, when they were sent to Gorky (now Nizhny Novgorod) for construction of the first two prototypes. The type's first flight was reportedly planned for 25 October 1994, and extensive testing of this large twin canard-delta combat aircraft included high-speed taxiing trials from December 1994. Despite the fact that the first two prototypes are complete at Zhukhovskii, and that the first one is actually taxiing, a strict ban on photography has remained in force. Defence Ministry approval for flight clearance was still awaited in mid-1995. The cause of the delays to the first flight have not been confirmed, and a variety of alleged problems have been mentioned. It has been said that funding had not been made available for high-altitude testbed clearance by TsIAM of the aircraft's Lyul'ka thrust-vectored engines, although other sources suggested that the prototypes would fly with standard non-vectoring AL-35s or with Kobchyenko R-79Ms (modified versions of the Yak-41's R-79 engine). The likely engine for any production 1-42 is the Saturn AL-41. Other rumours suggested that flight control software problems were responsible, or even that flight control actuators had not been delivered.

Two-seat 'Hokum' revealed

Russian army aviation is currently evaluating four of the 12 pre-series Ka-50s built to date, although Kamov General Designer Mikheyev's concept of a single-seat attack helicopter is reportedly meeting some resistance. Mikheyev revealed at the Paris air show that a two-seat version is now being developed. This follows the army's emphasis on all-weather and

night attack, and will allow competitive evaluations to be made against Mil's new Mi-28N.

The proposed Ka-52 will seat its crew side-by-side in a new, wider forward fuselage, and will have uprated twin Isotov TV3 powerplants, although 85 per cent of the Ka-50's airframe will stay unchanged. The Ka-52 crew will retain the Ka-50's original and unique Zvezda blade-jettison and K-37 extraction system for simultaneous emergency escape. They will also have additional systems to augment the original Leningrad Elektroavtomatika single-cockpit automated weapons avionics, for night operations.

These systems will include a Zenit SHKVAL day/night sight similar to that in the Su-25TK in the nose, alongside the original infra-red imager, and provision will be made for night-vision goggles. Full dual controls will be fitted for training use, although an operational mix of Ka-50s and -52s is foreseen for eventual tactical deployment. Mikheyev said that export versions of these helicopters would be fitted with French avionics and nav/attack systems, and would incorporate Igla-V AAMs.

He further claimed that the Ka-50 outperformed any other attack helicopter in the world, as did its 30-mm cannon (by a factor of five), and its Tula Vikhr laser-guided anti-tank missiles (in range and combat efficiency). Kamov planned to submit the Ka-52 for Sweden's upcoming attack helicopter requirement, although it has not yet gone beyond the mock-up stage.

'Hind' upgrade plans

Realisation of long-term Mil plans to modernise with new Mi-28 rotor systems and modernised avionics some of the 1,000 or more Mi-24 and export Mi-35 'Hind' heavy attack helicopters flying throughout the world was confirmed by the inclusion of a mock-up Mi-35M (Modifitsirovannyi) at the 1995 Paris air show. The Mi-35M is planned as a joint venture with Mil's associated Rostvertol production plant and Sextant Avionique and Thomson-CSF in France. The example at Paris was a non-airworthy mock-up, based on a real 'Hind-E' airframe. It was fitted with the Mi-28's five-bladed composite main rotor and X-type tail rotor, had cropped stub wings and dummy

Above: No. 20(R) Squadron now operates the two-seat Harrier T.Mk 10 (above), with slightly revised squadron markings. Behind this aircraft are GR.Mk 7s of No. 20(R) (with old-style marking), and No. 1(F) Squadron displaying the new fin-stripe.

Below: The Harrier T.Mk 4Ns of the Fleet Air Arm have been updated with Sea Harrier F/A.2 systems, and rechristened T.Mk 8. This aircraft of No. 899 Squadron Operational Evaluation Unit wears the same gloss black colour scheme as applied to the Hawk fleet.

weapons, but currently incorporates only a few of Sextant's new avionics and Thomson-TTDs mission systems equipment.

These are eventually planned to include a navigation computer with Pixyz RLG INS/GPS inputs. This will be integrated with a Chlio FLIR, mounted alongside the radio command system's chin fairing and will employ a dedicated collimated display. Twin head-down and eye-level cockpit-display screens will be provided for both the Mi-35M pilot in the rear and the front-seat weapons systems operator, the pilot also having a Sextant VH-100 head-up display. Provision is being made by Sextant for the addition of night-vision goggles.

The new avionics will be fully integrated with the existing weapons systems of the 'Hind', although the Mi-35M replaces the original 12.7-mm chin-turret machine-gun and 1,430 rounds with the twin-barrelled GSh-23-2 cannon associated with the Mi-24VP. It will also double the current capability to carry eight 9M114 (AT-6 'Spiral') ATMs, or related 9M120 or 9A-2200 ground-attack versions on abbreviated stub wings. Other Mi-34M weapon options include Igla V defensive AAMs. Shortening the stub wings of the 'Hind' removes the old endplate pylon, but forms part of a vital weight reduction campaign for the Mi-35M, which reduces its equipped empty weight from 8620 kg (19,004 lb) to 8050 kg (17,747 lb).

Other contributions come from the Mi-28's 300-kg (661-lb) lighter titanium main rotor head and composite blades, and by using a fixed instead of a

retracting tricycle landing gear, saving another 100 kg (220 lb). Normal take-off weight comes down from 11200 kg (24,692 lb) to 10900 kg (24,030 lb), raising the IGE hover ceiling by 1000 m (3,280 ft) to 3000 m (9,843 ft). Dynamic ceiling goes up from 4500 m (14,764 ft) to 5750 m (18,865 ft). Funding is now being sought by the M-34M group, mainly from private sources, to fly a prototype within 18 months to two years. Meanwhile, Rostvertol is continuing limited production of the fixed cannon-armed Mi-24P for Russian army aviation, and of the generally similar Mi-35P for export. It is also awaiting a production decision for the Mi-28N all-weather successor to the 'Hind', which is still in competition with the Ka-50/-52 for a Russian military order. Four Mi-28 prototypes have so far been built by Mil-Moscow, and some reports suggest that another one was built by Rostvertol.

A similar Mi-24/35 upgrade agreement involving Mi-28 dynamic components but with Israeli avionics and nav/attack systems has been concluded between the Mil Moscow Helicopter Plant and Israel Aircraft Industries.

New trainers on show

MiG and Yakovlev are battling for a potential advanced trainer contract to replace up to 1,000 Aero L-29 and L-39 trainers in CIS air force service. There is also wider interest in their MiG-AT and Yak-130 next-generation trainers in the export market; these aircraft made their public debut in prototype form at the 1995 Paris air show.

PZL's innovative PZL-230F Skorpion project has been reworked again into its third major configuration, with high bypass ratio Textron Lycoming LF507 turbofans mounted on the top of the aircraft, shielded from IR sensors by the outward-splayed fins. PZL is touting the Skorpion as a highly-agile, low-cost battlefield attack platform, with the replacement of the Polish Su-22 fleet a major ambition.

Neither had then flown, although were due to begin flight development in the following months after extensive ground testing. Flight qualification will lead to a fly-off to determine Russian air force preference in 1996-97, assuming the necessary programme funding can be found. MiG's relatively orthodox and straight-wing AT (Advanced Trainer) was rolled out on 18 May, and is being developed as a joint venture with French participation. Sextant Avionique is supplying most of the avionics systems for the 4610-7000-kg (10,163-15,432-lb) MiG-ATF (for French) version, which is also fitted with twin SNECMA 14.12-kN (3,175-lb) Larzac 04R20 turbofans. The prototype is fitted with Sextant's modular twin multi-function cockpit displays, a head-up display and associated nav/com equipment including Topstar GPS, Totem RLG INS, and a radio altimeter. A fly-by-wire control system is configurable to simulate the handling of next-generation combat aircraft.

A second prototype, the MiG-ATR, retains Russian avionics and is expected to retain the Larzac turbofans, although mutually acceptable terms for the engine programme had still to be

agreed with SNECMA this summer. A single-seat light-strike version, the MiG-ATB, is planned for export.

Composites are used for a large part of the MiG-AT airframe, including the now mid-set tail and control surfaces, flaps, rear fuselage, cowlings, intake trunks, under centre-section and forward-fuselage side panels. Three leading-edge slat sections on each wing extend with the flaps for optimum short-field performance. Boat-tail rear-fuselage split air brakes are of metal construction. Design maximum speed is 1000 km/h (540 kt) at the MiG-AT's 15500-m (50,800-ft) service ceiling, or 850 km/h (458 kt) at sea level.

A maximum speed performance of 1050 km/h (566 kt) is claimed for Yakovlev's more aerodynamically advanced Yak-130 trainer, which is optimised to reproduce the controllability and handling of such current types as the Su-27 and MiG-29 at angles of attack of up to 35°. It has a chined forward fuselage continuing from inner wing leading-edge extensions above the lateral air intakes, and a leading-edge sweep of 32°.

Paris '95 saw the Western debut of the Su-34, a bulky strike/attack derivative of the Su-27. The side-by-side accommodation is optimised for long-duration missions, a rudimentary galley and toilet facilities being provided behind the pilot and WSO. At Le Bourget, Sukhoi claimed this aircraft to be theSu-32FN, a maritime strike version of the Su-34. The aircraft in question is the third Su-34, representing the production version which followed two Su-27IB/KU development prototypes.

Powerplants are twin FADEC-controlled 21.6-kN (4,850-lb) thrust turbofans, developed and built by Povazske Strojarnye in Slovakia from the original Progress/Zaporozhye DV-2, in conjunction with Klimov and TsIAM. Production Yak-130s, possibly including a single-seat ground-attack version, will be powered by two uprated DV-2S or RD-35 turbofans.

With a normal take-off weight of 6200 kg (13,670 lb), including 1650 kg (3,638 lb) of internal fuel, the Yak-130 also features a computerised, programmable variable stability fly-by-wire control system developed from that of the Yak-41. Nav/attack systems include a HUD and head-down display in the deeply-stepped cockpits, plus other avionics from AlliedSignal.

Yak-130 construction is mainly metal, although composites are found in the tail section and rear-set stabilator. Dimensions of the Yak-130 include a span of 10.424 m (34.2 ft), a length of 11.3 m (37 ft), and a height of 4.7 m (15.5 ft). Both the MiG-AT and Yak-130 are fitted with Zvezda's new rocket-boosted K-93 lightweight ejection seats, offering zero-height and zero-speed escape capabilities for an equipped weight of only 58 kg (128 lb). Russian air force requirements alone are estimated at up to 800 aircraft, with a planned service entry date of around 1998.

UKRAINE:

Strategic bombers dispute

Hard and prolonged bargaining over financial terms was reported to have

resulted in an agreement earlier this year to return to Russia the 19 Tupolev Tu-160 'Blackjack' and 25 turboprop Tupolev Tu-95MS 'Bear H' strategic bombers based on Ukrainian territory at the time of the break-up of the USSR. Their 564 Kh-55 (AS-15 Kent) nuclear cruise missiles had been returned to Russia by Ukraine under an earlier agreement, which also included 416 other ALCMs and ICBM atomic warheads. The deal on the aircraft now appears to have foundered, with Ukrainian Defence Minister Valery Shmarov reporting a loss of interest in the deal by Russia, perhaps due to funding difficulties.

The Tu-160s of the 184th Heavy Bomber Regiment have been based at Priluki, in Ukraine, but have been mainly grounded through lack of spares. On their return to Russia they are expected to operate with the five Tu-160s of the VVS 196th Heavy Bomber Regiment at Engels air base, armed with extended-range RKV-500B (AS-15B Kent) ALCMs, as part of the national deterrent forces. No more Russian air force (VVS) orders have been received for Tu-160s, but a few are still being completed from earlier contracts at Kazan.

Most of Ukraine's Tu-95s have been operating with the 182nd Heavy Bomber Regiment at Uzin-Shepelovka, alongside the 409th Strategic Tanker Regiment with 20 Il-78 Midas air-refuelling tankers, and these are expected to return to Russia. Ukraine will continue to operate nearly 30 supersonic swing-wing Tu-22M 'Backfires', however, and was also left with about 55 older Tu-22 'Blinders'.

Middle East

BAHRAIN:

More F-16s sought

Bahrain's budget-limited plans to expand its force of 12 F-16C/Ds have resulted in negotiations for 18 F-16Ns withdrawn from US Navy aggressor training roles earlier this year. The F-16s would be exchanged for eight similarly low-houred F-5Es and four F-5Fs operated by the Bahrain since 1985. Although the F-16Ns are comparatively low-houred, they have high fatigue counts from their air combat training usage, and have suffered structural problems. They would require considerable upgrading to equip them to USAF F-16 standards but are much cheaper than the surplus USAF F-16A/Bs currently being offered for sale.

UAE:

Major procurement planned

Having received 36 Mirage 2000s from late 1989, the Abu Dhabi air force is evaluating possible successors in a programme which could be worth up to $5 billion.

Early deliveries of the Mirage 2000-5, and later the Rafale have been promised by France. Russia is another contender, with cut-price proposals for the Sukhoi Su-35. US proposals for the F-16U and the F-15U have been strengthened by Congressional approval of the AIM-120 AMRAAM for inclusion in the package. The selected aircraft will be required to operate with the UAE's own range of PGM-1A/B, -2, -3 and -4 Al Hakim stand-off ASMs, developed over several years by the former Ferranti International group (now part of GEC-Marconi Dynamics).

Britain's offered lease to Abu Dhabi of up to 24 surplus RAF Tornado GR.Mk 1s is an interim proposal, pending a final UAE decision on its new combat aircraft, for which the EF 2000 is a long-term contender.

Panther procurement

A $235 million contract has been signed by Abu Dhabi for seven Eurocopter AS 565SA Panthers for shore-based and frigate operation. These will have search radar and dipping sonar, and Aérospatiale AS15TT anti-ship missiles and homing torpedoes.

Also included is the upgrading by Aérospatiale with new radar and ASW equipment of Abu Dhabi's five remaining AS 532UC Cougars. AM39 Exocet anti-ship missiles will be included within the planned upgrade contract. This also covers modernisation of two ADAF AS 532UL Super Pumas used for VIP transport. Cougar redeliveries will start in mid-1998.

Transport and trainer plans

The Abu Dhabi air force has ordered 12 Grob G.115 all-composite aerobatic light trainers costing $5.5 million, with options on 12. A further requirement for up to four light maritime patrol aircraft favours the CN.235MPA.

Coastal defence missiles

Also in the UAE, Dubai has recently taken delivery from China of 30 CPMIEC HY-2 anti-ship missiles. Developed from the former USSR's SS-N-2 'Styx' AShM, the HY-2 is configured with small delta wings and an anhedral tailplane. It is ramp-launched using a ventral solid fuel booster, and has a range of about 30 km (16 nm).

With a launch weight of some 2440 kg (5,380 lb) (less booster), including a massive 513-kg (1,131-lb) armour-piercing warhead, it employs radio command guidance and monopulse radar terminal homing. It has also been supplied to other Middle Eastern countries, including Egypt, Syria and Iran.

One of several contenders for MiG-21 upgrade work, IAI's Lahav Division flew the first MiG-21-2000 after rework with Western equipment on 24 May 1995 (above). The photograph below shows six MiG-21s (including four MiG-21Us) undergoing rework at Lahav. The customer is alleged to be Cambodia. At the far end of the line is the MiG-23 which defected from Syria and which, wearing Israeli insignia, was extensively evaluated by the Israeli air force against Western types.

Southern Asia

INDIA:

More MiG-29s ordered

Agreement has been reached between the Indian and Russian governments for the purchase of another 30 MiG-29s, costing around $1.5 billion. These will supplement the survivors of 65 MiG-29 'Fulcrum-A' fighters and five two-seat MiG-29UB 'Fulcrum-B' combat trainers delivered between 1987-90 to equip three Indian air force squadrons, plus late 1994 orders for eight and two attrition replacements, respectively.

Since India is reportedly expecting all its additional MiG-29s before the end of 1995, they will be drawn from stocks of undelivered 'Fulcrum-As' or 'Fulcrum-Cs' in storage at Moscow's MAPO Lukhovitsky factory airfield. These were ordered but not paid for by the former Soviet air force, and are expected to undergo a modest avionics and weapons system upgrade to MiG-29S standard before delivery. It has been reported that plans are also being finalised with HAL to upgrade all the IAF's MiG-29s to MiG-33 (MiG-29M) standard, although such a conversion would not seem to be practically possible. If MiG-MAPO reports that MiG-33 development is being completed, India would be a likely export launch customer for the type, which would offer advantages of commonality with the existing MiG-29 fleet, while giving much greater multi-role capability.

Su-30MK for India?

India's new arms deal with Russia may include up to 48 two-seat Sukhoi Su-30MKs. Developed from the original two-seat Su-27PU, the MK is a multi-role version of the Su-30 long-range interceptor with provision for launching up to 8000 kg (17,637 lb) of PGMs with ranges up to 120 km (65 nm). Its retention of the basic 'Flanker-B' weapons system makes its air-to-ground capability rudimentary, and PGM compatibility seems unlikely without modifications similar to those which have resulted in the MiG-29SM. With a 1000-kg (2,204-lb) increase in maximum take-off weight to 34020 kg (75,000 lb) the MK does have an impressive endurance. Radar modifications give a dual target engagement capability, and presumably allow carriage of the new Vympel AAM-AE (R-77M/AA-12). An inflight-refuelling probe is fitted. Su-30MK deliveries are unlikely before early 1997.

MiG-21 developments

India has bought 10 MiG-21UMs from Bulgaria and Hungary to balance attrition in its 'Fishbed' fleet, and another 16 Jaguar ground-attack fighters have been ordered from HAL. India's MiG-21bis upgrade is still being finalised, however. Russia's upgraded MiG-21-93 prototype made its first flight on 25 May from the Sokol state-owned aircraft plant at Nizhny Novgorod, while negotiations for a major contract from India for conversions of up to 120 IAF MiG-21bis fighters were nearing their final stage. At that time, the MiG-MAPO group was still facing strong competition from Israel Aircraft Industries, Elisra, Elta and other upgrade specialists for this lucrative contract

The MiG-21-93's new weapons systems are based on installation of the 300-kg (663-lb) multi-mode Phazatron Kopyo pulse-Doppler fire-control radar. Also included are head-up, helmet and CRT displays, a new computer, INS, digital air data computer, and other advanced systems. External changes are mainly limited to a new intake shock-cone with an annular cooling bleed for the Kopyo installation, and a one-piece windscreen. Angular fairings above each wingroot, culminating in a double set of flare dispensers, accommodate 120 rounds of 26-mm IR decoys.

The Kopyo radar gives the MiG-21-93 a beyond-visual-range interception capability and compatibility with such advanced Russian AAMs as the Vympel R-27R (AA-10A 'Alamo') or RVV-AE or R-77M (AA-12 'Adder'), as well as the R-73E (AA-11 'Archer') or close-combat R-60 (AA-8 'Aphid'), plus a wide range of smart ground-attack missiles. Some Western avionics may also be installed in the MiG-21-93, including a Sextant Avionique ring laser-gyro INS with a GPS interface and a lightweight RWR, probably the Dassault Electronique EWS-A, with chaff and flare dispensers.

Further Do 228 orders

The IAF will place additional orders for the HAL-built Dornier 228, augmenting its fleet of 25 by 18. The IAF has two logistic support/communication squadrons (Nos 41 and 59) operating the Dornier 228s. They will be employed for multi-engine conversion training.

PAKISTAN:

More naval Alouettes

Four ex-RNAF SA 319B Alouette IIIs have been sold to the Pakistan navy, supplementing a similar number previously assembled locally at Dhamial. The ex-RNAF examples were delivered to Pakistan in February 1995, and two more were airlifted from Soesterberg in January to the Chad air force.

Far East

CAMBODIA:

MiG-21-2000 in flight test

A MiG-21bis fighter upgraded by Israel Aircraft Industries' Lahav Division made its initial flight at the adjacent Ben Gurion International Airport on 24 May. This was the first of 15 being modernised by IAI for the Cambodian air force under an $80 million contract. After the first 40-minute sortie, IAI's MiG-21-2000 made five more flights before the Paris air show to test the new avionics in all modes, at supersonic speeds.

Specific IAI upgrades for the MiG-21bis include a new MIL-STD1553 databus, with a new HUD and monochrome multi-function HDD; more accurate navigation and weapons release equipment linked with INS/GPS; a new air data computer; and a new armament interface unit. This automates release of Eastern or Western weapons, and control of radar. The MiG-21-2000 is also equipped with IAI's tactical datalink (JHAPS) which transfers operational data and location between groups of aircraft and ground stations. Other upgrades include a one-piece windshield,

Above: Modernisation of the Royal Thai air force's transport assets has been hastened by the arrival of the first Alenia G222. Six are on order.

Below: Former US Navy EA-7L (now a TA-7C) 156794 is the first of the Royal Thai navy's Corsair IIs to be delivered. This aircraft was delivered to Thailand, from NAS Jacksonville on 13 May, but stopped over at NAS Meridian where future Thai A-7 crews are undergoing training with CTW-1.

a new stick grip for HOTAS operation, and a redesigned cockpit panel with new avionics and western nav/com equipment.

Lahav is prepared to upgrade the existing RP-22/S-21 Saphir (NATO Jay Bird) radar in the nosecone of late production MiG-21s, to present improved processor and video outputs to an El-Op HUD and multi-function displays, or will install an Elta EL/M 2032 radar as a higher-cost option. IAI subsidiaries are also supplying systems for other MiG-21 upgrades.

CHINA:

More Su-27s ordered

Following deliveries of 24 Sukhoi Su-27s and two Su-27UBs from a 1991 order, China has concluded a follow-up contract with the Sukhoi OKB. Apart from a second batch of 22 Su-27s and two Su-27UBs, an MoU is also included for a 'Flanker' production licence – a quantum leap in technology for the Chinese aerospace industry.

New Sino/Pakistani/ Russian fighter revealed

A new lightweight fighter designated FC-1 (Fighter, China) was unveiled at the 1995 Paris air show by the China National Aero Technology Import & Export Corporation (CATIC). The Chengdu Aircraft Corporation has been working on the FC-1 since 1991, as a replacement for the MiG-21 and its planned Super 7 successor, through a joint programme with MiG-MAPO

in Moscow and the Pakistan Aeronautical Complex at Kamra. Russia's Klimov Design Bureau is also supplying the 81.4-kN (18,300-lb) thrust RD-33 turbofan, in the form of a planned RD-93 version.

Shown in model form at Le Bourget, the FC-1 resembles a cross between a MiG-21 and a Saab JAS 39 Gripen. It also bears a resemblance to the stillborn Mikoyan lightweight fighter abandoned during the mid-1980s, and provisionally given the MiG-33 designation now taken over by the MiG-29M. This aircraft was abandoned because it offered unnecessary competition to the MiG-29 without any major advantages and because it did not conform to the twin-engined doctrine belatedly but enthusiastically embraced by the design bureau. The FC-1 differs from the original 'MiG-33' primarily in its use of lateral Gripen-type intakes rather than a single chin intake.

Design features of the FC-1 include extended, notched LERX fairings with sufficient internal volume for fuel stowage, a cropped-Delta active wing, a large nose radome, and four underwing, one centre-line and two wing tip weapons stations. Armament includes a nose-mounted twin-barrelled Gsh-23-2 cannon of Russian origin. Pakistan will select its own avionics and weapons systems from Western sources for its version of the FC-1, while the Chinese air force (AF/PLA) will use indigenous systems and equipment.

With a planned maximum speed of around Mach 1.8, the FC-1 will have a similar performance to the F-16, but will use conventional hydraulic servo-operated flight controls with an analog

fly-by-wire back-up. FBW could be used as the primary system at a later stage. Three flight prototypes and two static airframes of mainly conventional metal construction are currently being built at Chengdu, plus some components in Pakistan, for a first flight date in early 1997. MiG-MAPO said in Paris that the FC-1 project was currently about 40 per cent completed. A final assembly line will also be set up in Pakistan to deliver the first FC-1s by about 1999. Those for the Pakistan air force will be fitted with Martin-Baker Mk 10 zero-zero rocket-boosted ejection seats as standardised by the PAF for its earlier Chinese-supplied F-6s and F-7s. AF/PLA FC-1s will probably be equipped with developments of China's current HTY-4 ejection seats, or with exported Zvezda K-36 seats. Pakistan will co-operate with CATIC in achieving export sales of the FC-1 and target fly-away unit cost has been set at an optimistic $10 million.

K-8 trainer progress

Some 22-25 K-8 jet trainers are reported by the Nanchang Aircraft Manufacturing Company (NAMC) as being built in batch production, following delivery of the first six last year to Pakistan. Satisfied with its evaluation of the programme, the PAF is finalising a larger follow-up order of K-8s, which it calls the Karakoram, a total requirement for 75 to replace T-37s at the Risalpur Air Academy.

The AF/PLA is expecting its first K-8s, also with Garrett TFE731 turbofans and licence-built Rockwell Collins avionics, later this year. These will have indigenous HTY-4 ejection seats rather than the Martin-Baker Mk 10 zero-zero units specified by the PAF, plus other non-Western equipment.

INDONESIA:

F-5 upgrade contract placed

Belgium's SABCA group has been awarded a $40 million contract to upgrade eight Northrop F-5Es and four F-5Fs of the Indonesian air force (TNI-AU) with new nav/attack systems. After prototype installations in Belgium, the remaining TNI-AU aircraft will be modified in Indonesia.

KOREA:

First Orion deliveries

In early April Lockheed Martin and the USN delivered the first two of eight P-3C Update III Orion maritime patrol aircraft ordered by the South Korean government in 1992. After training with the USN's VP-30 at NAS Jacksonville, FL, RoKN crews took part in the delivery flights from Lockheed Martin's Marietta factory, in Georgia. All eight P-3Cs will be delivered this year. Samsung Aerospace has recently been awarded a $5 million three-year repair and overhaul contract for 20 Allison T56 turboprop engines from Dutch navy Lockheed P-3Cs.

Ka-32 helicopters received

Four Kamov Ka-32Ts have been accepted by South Korea, as partial settlement of a massive $1.47 billion trade deficit. They have a nominal unit cost of about $2 million each, and four more are expected in July as further offsets. Their operating agency has not yet been specified.

New KTX-1 prototype

Daewoo was planning to fly a third prototype KTX-1 turboprop tandem trainer in July, with an uprated Pratt & Whitney Canada PT6A-62 engine. This develops 950 shp (709 kW) rather than the 550 shp (410 kW) of the PT6 version in the two previous prototypes, which began their flight development programme in 1992. The RoKAF is expected to take delivery of the first of about 100 planned production TX-1s in 1998, to replace current Cessna T-41s and T-37s.

MALAYSIA:

New RMAF MiG-29 units

Two new units are being formed by the Royal Malaysian air force to operate the 16 MiG-29Ns and two MiG-29NUBs delivered in pairs from 21 April. These were purchased in a reported $560 million contract (from a total package cost of $1.6 billion). The aircraft were delivered by Volga-Dnepr An-124s for assembly at Kuantan. MiG-29 operational conversion and ground training is being undertaken by No. 17 Sqn, alongside No. 19 (Air Defence) Sqn at the same base.

Although described in some circles as being basically similar to the export MiG-29SE or, more precisely, -29SM, the MiG-29N is actually a 'Fulcrum-A' version and is regarded by the bureau as a MiG-29SD. The aircraft features an upgraded Phazatron NO19M Topaz radar conferring TV-guided ASM capability, as well as operation of the medium-range Vympel R-77 RVV-AE (AA-12 'Adder') active radar-homing AAM. The MiG-29N designation is reported to be a local one, and is not recognised by MiG-MAPO.

The MiG-29N is the first 'Fulcrum' to have a longer-life RD-33-3 turbofan. Developed by the Chernyshov/Omsk engine enterprise for Malaysian needs, the RD-33-3 has been designed to extend the original 700-hour service life initially to 2,000 hours and eventually up to about 4,000 hours, with increased reliability and lower lifecycle maintenance costs.

This has been achieved by strengthening the fan rotor, HP compressor assembly and three-stage labyrinth seals of the combustion chamber, plus the installation of more durable bearings. The main engine parameters are also claimed to have been stabilised, including a maximum dry thrust of 50 kN (11,240 lb), and full afterburning output of 81.39 kN (18,298 lb) with an s.f.c. of 0.77, for a dry weight of 1055 kg

(2,326 lb). In the MiG-29, the RD-33 has been cleared for operations up to Mach 2.35 (1500 km/h; 809 kt) IAS at 11000 m (36,089 ft).

MiG-MAPO reported some $1.5 billion in turnover from 60 MiG-29 export deliveries in 1993-94 to Hungary, Iran, Slovakia and Romania, plus subsequent sales to India and Malaysia. About 400 MiG-29s are now claimed by MiG-MAPO to have been exported to 22 countries, from total sales of over 1,200. Current fly-away export price is quoted as around $25 million. A briefing on the latest Fulcrum export versions will appear in a forthcoming issue of *World Air Power Journal*.

TAIWAN:

F-5 upgrade challenged

Government plans for the upgrade and conversion of 10 F-5Es to RF-5E TigerEye standards by Singapore Aerospace at a cost of some $50 million have been challenged by Taiwan's own aerospace industry because of its current low workloads. These result from purchases of F-16s and Mirage 2000-5s for the RoCAF, leading to cuts in procurement of Taiwan's Ching-Kuo.

Under the terms of the proposed contract with Singapore, Taiwanese industry stands to receive only about $15 million in direct offset work, although this would be increased if further upgrades of the RoCAF's 250 or so remaining F-5s were to follow. Singapore Aerospace, in conjunction with Northrop, had previously modified eight F-5Es to RF-E standards for the RSAF, with extended noses containing six apertures and combination mounts for up to four vertical- and oblique-mounted 70-mm cameras.

THAILAND:

P-3 Orion deliveries

Thailand became the 15th Orion operator when the Royal Thai navy's 101 Squadron took delivery of the first of two refurbished P-3T ASW aircraft on 16 February at U-Tapao. The RTN's third and last Orion, a UP-3T, was due for delivery by November.

VIETNAM:

Su-27s delivered

The Vietnamese People's Army air force has become the second export customer for the Sukhoi Su-27, with at least two 'Flankers' delivered earlier this year. They are supplementing several squadrons of MiG-21bis and Su-22s, for which upgrade programmes are now being discussed.

Above: On 2 May 1995 Fuerza Aérea de Chile took delivery of a Boeing 707 equipped with the IAI/Elta Phalcon AEW system. The aircraft is known locally as the Condor.

Below: The AH-64D is undergoing AAM compatability trials in the USA. Missiles, such as the US Stinger or the UK's Starstreak will be housed in a new box launcher.

Australasia

AUSTRALIA:

C-130J Hercules procurement plans

In June, Lockheed Martin responded to requests for quotations by the Royal Australian Air Force, which plans to become the second export customer for the Lockheed C-130J (after the RAF). Funding had already been allocated by the Australian Defence Department to acquire 12 C-130J-30s to replace the ageing C-130Es of No. 37 Squadron from early 1997.

Several Australian companies are receiving component production contracts as part of the offsets included in the RAAF's initial C-130J order. The prototype C-130J is scheduled to make its first flight in December 1995, with RAAF deliveries expected to start in late 1997. The fly-away unit cost is quoted by Lockheed as $40 million.

Australia's 12 existing C-130Es date from 1966, though the Air Force operated C-130As from 1958 until the mid-1980s. It had once been hoped that the C-130Es would have been superseded

by C-130Hs, by the mid-1990s, but this plan was shelved due to lack of funding. Australia has also been touted as a prime customer for any C-130-based AEW&C aircraft, but attention is now being focused on smaller types such as the Saab Erieye.

Caribou replacements evaluated

About 15 smaller twin-turboprop transport types are being evaluated as replacements for the RAAF's DHC-4 Caribous, for which the CASA CN.235M9 or IPTN CN.235-330 and Alenia G222 are prime contenders. Now known as the Phoenix, IPTN's CN.235 is increasingly favoured, following its backing in Australia by Hawker de Havilland and Honeywell. The RAAF recently took delivery of four Boeing CH-47D Chinook medium-lift helicopters. These were upgraded from 11 ex-RAAF CH-47Cs traded in to Boeing in 1993, which retained the remainder for similar modernisation and resale as payment.

South America

ARGENTINA:

US A-4 supply finalised

The long-planned (and at time politically stormy) US supply of 36 surplus McDonnell Douglas A-4M Skyhawks to Argentina, and their upgrading through Lockheed Martin auspices in a $480 million FMS contract, was finalised in June. Some $200 million of this contract is allocated for the management over a 25-year period of Argentina's formerly state-owned Fabrica Militar de Aviones production and overhaul facility at Cordoba by Lockheed Martin Air Services, through its new Lockheed Aircraft Argentina subsidiary.

LMAS is also prime contractor for the upgrade of the first 18 FAA A-4Ms by the installation of Westinghouse APG-66/ARG-1 multi-mode radar, new cockpit displays and nav/attack systems, at its Ontario, CA, factory. It will also supervise similar modifications of the second batch of A-4Ms in Argentina by FMA, through the $280 million upgrade contract.

CHILE:

AEW Phalcon delivered

Formal Chilean air force acceptance of the Boeing 707 converted to Phalcon standard took place on 2 May . This involved installation of lateral conformal Elta EL2075 phased-array radar antenna packs and a nose radome, plus associated processing equipment for airborne early warning roles. Although work on the Phalcon through a $150 million IAI contract was completed in early 1995, Chile delayed acceptance pending the resolution of mission system software problems. IAI technicians are helping with continuing software upgrades during Phalcon's initial service.

Hawker Hunter retired

Chile finally retired the Hawker Hunter from front-line operational service in April 1995, mounting a ceremonial flypast using two two-seaters and two single-seaters. Pilots for this flypast included the FACh C-in-C, General

del Aire Ramon Vega Hidalgo, the Chief of the Air Staff, General de Aviacion Fernando Rojas Vender, and the last CO of Grupo 8, Comandante de Grupo Patricio Saavedra. Other pilots on the flight were General de Aviacion Jaime Estay Viveros, General de Brigada Florencio Duble and General de Aviacion Enrique Montealegre. Some 16 Hunters will reportedly be preserved (of which some may be allocated to foreign museums). Chile will reportedly retain two Hunters in airworthy condition, although they will no longer be regarded as front-line aircraft. With the ongoing dispersal and disposal of Switzerland's Hunters and the recent retirements in Oman and Singapore, this leaves only the Royal Air Force and Zimbabwe air force as active Hunter operators.

Above: US Navy Boeing E-6 TACAMO aircraft of the Tinker-based Strategic Communications Wing One (VQ-3 and VQ-4) are being fitted with the MILSTAR satcomms hump. The fleet is due to take over the 'Looking Glass' airborne command post mission in 1997, with the resultant retirement of the remaining Boeing EC-135Cs operated by the 55th Wing.

Below: The 'Sluggers' of VF-103 have adopted new markings for their F-14Bs. The squadron is the only Tomcat unit within Air Wing 17.

North America

UNITED STATES:

F-111 for early retirement

The USAF has finalised plans to retire the F-111E and F. The retirement will take place earlier than planned, and will free funds for new projects. The Air Force has about 100 F-111s assigned in four squadrons, and these will commence withdrawal from Fiscal Year 1996. The 27th FW will not inactivate and will transition to the F-16C/D (with 54 aircraft assigned). It is not clear where these aircraft will come from, but they may be from the final production purchased during FY 1993 and 1994.

The three dozen remaining EF-111As are not affected, and their retirement is still scheduled between 1997 and 1999. They may receive a new lease on life after the importance of EW was demonstrated by the shoot-down of an F-16 in Bosnia.

On 14 March 1995, an EF-111 of the 46th Test Wing at Eglin AFB, Florida made the first flight of an aircraft modified under the Grumman Systems Improvement Program (SIP) which provides new sensors and avionics. At the same time the administration was pressing its plan to retire the EF-111 from service – which would make the SIP upgrade irrelevant.

The flight test programme of the first SIP-configured EF-111 was scheduled for completion in July 1995. It will test an encoder/processor which increases the number of hostile radars the AN/ALQ-99E suite can cope

with. No decision has been made, and no funds allotted, to proceed with SIP for the remainder of the EF-111 fleet.

The USAF still wants to put the EF-111 out to pasture and, with it, the SIP programme. The service has 40 EF-111s in inventory, 27 kept combat-ready. All EF-111s are concentrated at Cannon Air Force Base. The 429th Electronic Combat Squadron is one of the 'most deployed' USAF units. In 1994, EF-111 crews spent up to 155 days per year away from home, cited as a morale problem. The figure has since dropped to 132 days but still exceeds the USAF's 'maximum desirable' 120.

Prior to the much-publicised F-16 shoot-down, the future of the EF-111 and other US electronic warfare aircraft was to have been decided by a Joint Tactical Aircraft Electronic Warfare Study (JTAEWS) scheduled for completion in July 1995. Retiring the EF-111 would force the Pentagon to rely on the Navy's EA-6B for radar-jamming. Whatever the JTAEWS recommends, the Air Force is bucking Congress and pressing ahead with plans to phase out the EF-111 Raven fleet between 1997 and 1999.

Critics argue that the EA-6B Prowler is too slow to keep up with modern strike packages and lacks the range of the EF-111. Further, there is dispute as to whether the Navy's current fleet of 127 EA-6Bs is enough for its own requirements. In the aftermath of the loss of an F-16 to a radar-guided SA-6 missile, the retirement of the EF-111 is being argued again and seems certain to be postponed.

C-17 now operational

The 17th Airlift Squadron at Charleston AFB, South Carolina was declared operational on 17 January 1995. General Robert Rutherford, the C-in-C of Air Mobility Command, announced that the squadron had attained Initial Operational Capability (IOC). The declaration enables the C-17 to be used on any type of military or humanitarian mission. The squadron will now make preparations for a 30-day reliability, maintenance and availability evaluation set for July 1995. This intensive evaluation will be conducted under peacetime and simulated wartime emergency conditions. In the meantime, the 14th Airlift Squadron began conversion from the C-141B to the C-17A. The squadron should receive its complement at a rate of approximately one per month, with the 12th example due early in 1996.

The Globemaster is undergoing minor modifications to clear it to drop paratroopers after separation problems were discovered during tests at Yuma Proving Ground, Arizona. The tests showed risks in simultaneously dropping troopers from doors on opposite sides of the aircraft. The C-17 is expected to drop 102 paratroopers – 51 from a door on each side – in 51 seconds.

Meanwhile, a C-17 operated by the 17th Airlift Squadron, Charleston AFB, South Carolina made a 2 March 1995 flight to Travis AFB, California, Elmendorf AFB, Alaska and Yokota AB, Japan, completing the type's first trans-Pacific mission on 6 March. The Pentagon is not committed to purchase more than 40 C-17s (down from 120) but the real figure is expected to be 50 to 70.

Extenders move base

The fleet of 59 KC-10A Extenders began moving to new bases during September 1994 as part of a reorganisation of tanker assets. The 60th Airlift Wing at Travis AFB, California was redesignated an Air Mobility Wing on 30 September, while the 438th Airlift Wing at McGuire AFB, New Jersey inactivated on 30 September to be replaced by the 305th Air Mobility Wing the following day. The two wings had received their first KC-10s shortly before the change of designation, with the 9th Air Refueling Squadron moving to Travis AFB from March AFB, California on 1 September. The 6th ARS began relocating from March to Travis during the spring of 1995, completing the West Coast part of the reorganisation. The first 10 Extenders to be based at Travis were formerly with the 4th Wing at Seymour Johnson AFB, North Carolina. To enable operations at March AFB to continue during the transition period, the 722nd ARW was formed when the 22nd ARW was transferred to McConnell AFB, Kansas on 1 January 1994. This is only a temporary measure, with the KC-10s due to complete relocation by September 1995. The 32nd ARS moved from Barksdale AFB, Louisiana to McGuire AFB during September 1994, with the 2nd ARS beginning its move during the late spring of 1995.

A small number of KC-10As, possibly as few as six, are due to be assigned to the 22nd ARS at Mountain Home AFB, Idaho during 1995, replacing a similar quantity of KC-135Rs. These will be the only Extenders assigned to Air Combat Command. The remaining 53 will be divided between the two Air Mobility Wings of Air Mobility Command.

OL-UK U-2s move base

The three U-2s of Operating Location – United Kingdom (OL-UK) moved from RAF Alconbury to RAF Fairford on 15 March 1995. The operation has remained in Europe to perform reconnaissance duties in support of the 'No-Fly Zone' over Bosnia. They are fitted with a variety of sensors and equipment. One has the Senior Span satellite communications system in a large dorsal teardrop fairing. The other two U-2s house their equipment in the standard nose, Q-bay and super pods, and include the ASARS-2 system in an extended nosecone. The super pods contain a variety of sideways-looking radar sensors ands Sigint antennas. The move to Fairford will only be temporary as the unit is due to relocate to Italy by the beginning of Fiscal Year 1996.

USAFE involvement in UN operations to be reduced

New Air Force Chief of Staff General Ronald R. Fogleman announced new initiatives aimed at relieving the pressure on USAFE units supporting UN peacekeeping efforts. In recent months Air National Guard and Air Force Reserve squadrons have deployed to Incirlik for Operation Provide Comfort, and to Aviano for Operation Deny Flight. Reservist units deploy to Europe for limited periods to enable active-duty squadrons to stand down from performing daily missions in potential war zones. Among the units participating in UN operations recently were the 466th FS, 419th FW, an AFRes unit from Hill AFB, Utah which staged to Incirlik with 11 F-16Cs. The reserve F-16s remained for two months with crews swapped at two-weekly intervals. The deployment was a joint arrangement with the 302nd FS, 944th FW from Luke AFB, Arizona which manned the Hill aircraft for the second half of the deployment. Incirlik also had six Air National Guard F-15As in residence for two months with aircraft and crews from the 122nd FS at NAS New Orleans, Louisiana and the 199th FS at Hickam AFB, Hawaii. The AFRes representation replaced the 52nd FW F-16Cs, while the ANG participants supplanted the 493rd FS from Lakenheath.

The F-15Es of the 48th FW were also given a four-month break from Operation Deny Flight at Aviano when Spanish air force EF-18 Hornets took over from early December. A mix of Air National Guard A-10 squadrons spent two months at Aviano from early December, enabling the 81st FS at Spangdahlem to enjoy a well-earned break. According to General Fogleman, the reserves will be more involved in both theatres in future.

Subsequently, General Fogleman highlighted the forces in Alaska as being earmarked for European operations, with the 54th Fighter Squadron from Elmendorf AFS deploying to Incirlik AB, Turkey on 13 April with six F-15Cs. The aircraft routed via Langley AFS, Virginia and RAF Lakenheath. Spangdahlem's base commander, Brigadier-General John Dallager, recently brought to the attention of Defense Secretary William Perry the domestic problems among his personnel attributed to the high amount of time they spend away from home. All four 52nd FW squadrons routinely deploy to Incirlik for Deny Flight, while in addition A-10 crews of the 81st and F-15 crews of the 53rd FS regularly perform sorties from Aviano for Deny Flight.

To alleviate the problem, the Air Force agreed in January 1995 to a 22-month deployment schedule which will see the burden shared more evenly. Apart from the PACAF F-15Cs, April saw a dozen F-16Cs from three ANG squadrons transit Spangdahlem for Incirlik, enabling the 23rd FS to stand

Among the rivals vying for lucrative F-5 upgrade contracts is the original manufacturer. In late April Northrop began flight tests from Edwards AFB of its F-5E Tiger IV demonstrator. This is fitted with new cockpit systems, Westinghouse APG-66 radar and other improvements aimed at vastly increasing the F-5's combat potential.

down from the commitment. The 120th FS from Buckley ANGB, 138th FS from Syracuse, and the 160th FS from Montgomery, each provided four F-16Cs, almost certainly for three months with pilots and ground crews rotated monthly.

USAFE has an eventual goal of each squadron performing no more than 90 days' annual deployment on military operations, with an additional 30 days on Flag-type exercises in the USA.

USAF unit news

The 70th Fighter Squadron at Moody AFB, Georgia received its first A-10As on 9 January 1995. The 347th Wing at Moody is the third and latest unit to gain composite wing status with the assignment of the A-10A alongside the 52nd Airlift Squadron with the C-130E, and the 68th and 69th Fighter Squadrons flying the F-16C/Ds.

The A-10 presence at Al Jaber Air Base, Kuwait was exchanged during January 1995 with the return of 24 aircraft of the 23rd Wing which deployed during the autumn of 1994. They were replaced by a similar number of A-10s from the 355th Wing at Davis-Monthan AFB.

The 53rd Fighter Squadron at Spangdahlem commenced a six-week rotational period of alert duty on 3 January 1995 to defend the central region of NATO. The squadron flew a training mission over the former East Germany, the first occasion that USAF fighters had been given unrestricted access to this airspace.

The 37th Airlift Squadron at Ramstein AB, Germany began applying an 'RS' tailcode to its C-130Es as aircraft were rotated through maintenance. In addition, the aircraft will feature a yellow and blue checkered tailband. The C-130Es are part of the 86th Airlift Wing, which also operates the C-9As within the 75th AS. The 76th AS is assigned the C-20A, C-21A and a CT-43A to perform medevac and VIP/communications duties, which will not

receive the tailcode. One of the Learjets has received a new VIP-style livery similar to that applied to the three Gulfstreams.

The Air Force Reserve anticipates extending its associate programme with the establishment of at least one squadron at Tinker AFB, Oklahoma to provide personnel to operate the E-3 Sentry. The AWACS community spends a great deal of time deployed overseas, with aircraft and crews performing 90-day rotations to Incirlik AB, Turkey and Riyadh, Saudi Arabia to enforce United Nations exclusion zones. The 1996 defence budget proposals include $25 million for funding to cover the cost of increased deployments by reserve forces.

New homes for old F-16A/Bs

The Air Force Reserve and Air National Guard have replaced many of their F-16A/B models with the F-16C/D, enabling the former versions to be retired to AMARC or sold to overseas customers. At least 50 were supplied to Israel during 1994. There are more than 250 F-16A/Bs in store, with several nations showing interest in obtaining them. Among those who have approached the US government for preliminary talks are the Czech Republic, Hungary and Poland, who wish to replace or supplement their MiG-21s. Poland has already made an approach for an unspecified number of F-16s. A US delegation visited Minsk-

Mazowiecki, near Warsaw, recently to check facilities at Poland's sole MiG-29 base. The Poles would like to replace the MiG-21 'Fishbeds' and MiG-29 'Fulcrums' of the 1 PLM at Minsk with the F-16. The displaced 'Fulcrums' would be transferred to the 28th PLM at Slupsk, enabling MiG-23s to be retired.

Iceland Eagle changes

Air Forces Iceland gradually lost its F-15C/D Eagles during 1994 when the 57th FS returned its aircraft to the USA. The commitment to defend the region and maintain a credible response for the Iceland/Faroes Gap has continued, with F-15s from Stateside-based units rotating to Keflavik. The 1st Fighter Wing from Langley AFB, Virginia sent five F-15Cs to Keflavik on 7 January 1995 to commence alert duties for a first 30-day period, with the commitment being rotated between the three squadrons at Langley. At the completion of the duty by the 1st FW, the alert commitment was transferred to the Air National Guard, and the 101st FS from Otis ANGB, Massachusetts. The only other ANG F-15 unit dedicated to the air defence role is the 142nd FS at Portland International Airport, Oregon.

F/A-18 Hornets are now widely used as Navy adversary aircraft, the F-16N fleet having been retired. A relative newcomer to the Hornet world is Key West-based VF-45.

In trials the F/A-18 has demonstrated an ability to carry no fewer than 10 Hughes AIM-120 AMRAAM missiles (in addition to two wingtip Sidewinders), representing the theoretical maximum combat persistence load. As well as the two fuselage stations, the missiles are carried on dual-rail launchers on each of the four wing pylons.

Three ANG Eagle squadrons have a tactical fighter role, but will add this function. The three squadrons are the 110th FS at Lambert Field, Missouri, the 122nd FS at NAS New Orleans, Louisiana, and the 128th FS at Dobbins AFB, Georgia.

E-6 Mercury

The Navy recently accepted the first E-6B Mercury fitted with the MIL-STAR satellite communications system housed in a large fairing aft of the cockpit. The first aircraft – 164406 – was completed during June 1994. Test work was carried out at Patuxent River and was completed in February 1995. The conversion was carried out by Chrysler Technologies Airborne Systems Inc. at Waco, with the remainder expected to be similarly modified over a two-year period. On completion, the E-6 will take over the airborne command post role from Air Force EC-135s at Offutt AFB.

The E-6s are based at Tinker AFB, Oklahoma and assigned to Sea Control Wing One (SCW-1), with crews assigned to VQ-3 and VQ-4 for Pacific and Atlantic Fleet operations, respectively. The selection of Tinker AFB to base the E-6s was to co-locate them with the USAF's similarly tasked E-3A Sentries, and to centralise both squadrons at one location suitable for both to reach their operating area. VQ-4 maintains a detachment at NAS Patuxent River, Maryland for the Atlantic Fleet, while VQ-3 detaches aircraft to Travis AFB, California for its Pacific Fleet operations.

The Navy has 16 E-6s allocated serials within four batches: 162782 to 162784, 163918 to 163920, 164386 to 164388, and 164404 to 164410. The aircraft are divided roughly between the two squadrons, with VQ-3 originally allocated eight airframes (162783, 162784, 163918, 163919, 163920, 164386, 164387 and 164388), while VQ-4 had just six (164404, 164405, 164407, 164408, 164409 and 164410). E-6A 162782 has been retained by Boeing for various trials and upgrades, while 164406 was the E-6B prototype.

Pacific Tomcats on the move

NAS Miramar, California has been the home of the Pacific Fleet F-14 Tomcat squadrons since the type commenced delivery in 1973. However, the Navy has announced plans for Marine Corps units based at MCAS El Toro, California to relocate to Miramar, and resident Navy squadrons will have to move elsewhere. The F-14 squadrons were to have relocated to NAS Lemoore, while the Navy reserve aggressor aircraft and the 'Top Gun' school would have moved to NAS Fallon, Nevada. The fate of the E-2 community was not made clear, although a move south to NAS North Island could be a possibility. The placement of the F-14 squadrons at Lemoore would not create an over-

crowding problem as several Tomcat units have been disestablished recently while there are fewer Hornet squadrons in residence at Lemoore due to the drawdown.

The Marine Corps F/A-18s began moving to Miramar in August 1994, with three and possibly four squadrons having taken up residence at their new base by April 1995. The F-14s do not appear to have started their move, and there seems to be some doubt about the plan as the Navy has voiced its opposition. The Navy secretary recently suggested that the Pacific Fleet F-14s should amalgamate at NAS Oceana, Virginia along with the Atlantic Fleet F/A-18s, thereby creating the ultimate super base. VF-124, the West Coast FRS, was disestablished in mid-1994, with all fleet replacement training now being centralised at Oceana under VF-101, although the latter has a permanent detachment at NAS Miramar. All of VF-101's aircraft display tailcode 'AD', with the F-14s stationed on the East Coast having nose modexes in the series 100, while those of the Miramar detachment are allocated modexes from 200. VF-101 aircraft from Oceana regularly deploy to NAF El Centro, California for weapons training on the nearby ranges. An interesting visitor at El Centro on 1 May was VF-101's T-34C 162299 (coded AD-003), which was painted white overall with a red upper section on the fuselage and tail and with a red-and-white checkered fintip. The aircraft is based at Oceana and must have taken many hours to have flown the 2,600 miles (4185 km) to El Centro. The aircraft landed shortly after the last of eight F-14Bs arrived.

JPATS selection

The Raytheon Beech Mk 2 turboprop trainer has been selected as the JPATS (Joint Primary Aircraft Trainer System) to replace USAF T-37 Tweet and USN T-34C Turbo-Mentor trainers. The aircraft is an Americanised version of the Pilatus PC-9 Mk II and is powered by a Pratt & Whitney PT6A-62 turboprop engine. The choice was a surprise to some. The JPATS competition had been thought to favour turbofan aircraft, with the Cessna CitationJet and Rockwell/Deutsche Aerospace Ranger 2000 being widely rumoured as the preferred aircraft.

The first batch of JPATS aircraft will total 141 (long term total of 711: 372 for the USAF and 339 for the US Navy) but will be only a fraction of potential sales for Raytheon's Beech division. The manufacturer is authorised to seek foreign customers and is expected to find interest in Greece, Mexico, South Korea, Thailand and Turkey. JPATS would be ideal for Korean requirements if the planned indigenous KTX-1 trainer fails to materialise. Foreign sales of the JPATS aircraft become important now that the Pentagon has 'stretched' the programme from 12 to 20 years, reducing planned production rates below the point where the builder can make a profit in the early years.

RAH-66 Comanche

The first RAH-66 Comanche helicopter for the US Army was rolled out at Sikorsky's Stratford, Connecticut facility on 25 May 1995 and was demonstrated (without having yet flown) at the Pentagon Building's helipad on 19 June 1995. The US Army hopes to acquire 1,292 Comanches.

Following its appearance at the Pentagon, the prototype was to be trucked to Sikorsky's West Palm Beach, Florida facility where its mission computer was to be installed. First flight was scheduled for November 1995. A second prototype was scheduled to begin flying the Comanche's reconnaissance mission equipment package in September 1998. The two RAH-66 prototypes are scheduled for seven years of flight testing that will run until 2002.

The RAH-66 is powered by two 925-shp (690-kW) LHTEC T800 turboshaft engines driving 39.04-ft (11.90-m) five-bladed, bearingless composite main rotor and 4.49-ft (1.37-m) 'fantail' anti-torque tail rotor. After difficulties staying within its design weight, the US Army now says that the RAH-66 is within the specified empty weight of 7,977 lb (3,626 kg).

When Pentagon officials proposed a major funding cut and indefinite production deferral for the Comanche earlier in the year the Army Staff came up with a novel idea to keep its top-priority programme alive. The Army devised a way to build an additional six EOC (early operational capability) aircraft within the programme's available funding. The EOC helicopters will be used for field evaluations and will protect a production option. The Army then won Pentagon approval of the restructured programme which will bring the Comanche to IOC (initial operating capability) in 2006. An option has been preserved for a 'speed-up' which would restore the one-time IOC target of 2003. All six EOC aircraft are to be delivered in 2001.

An unusual weapon for a Tomcat is the GBU-24 Paveway III 2,000-lb laser-guided bomb with BLU-109 penetrating warhead. The Tomcat can launch the weapon, but does not, as of now, have any autonomous designation capability.

F-16 shoot-down

An F-16C fighter of the USAF's 555th Fighter Squadron 'Triple Nickel', part of the 31st Fighter Wing at Aviano, Italy, was shot down by an SA-6 missile over Bosnia on 2 June 1995. Pilot Captain Scott F. O'Grady ejected and evaded capture for six days, part of that time with hostile Bosnian Serb troops passing within arm's length and shooting wildly. O'Grady established contact on his PRC-112 survival radio and directed his rescue. A Marine Corps TRAP (Tactical Recovery of Aircrew Personnel) team from the assault ship USS *Kearsarge* (LHD-3) – trained to fight at night – went in after daylight on 8 June 1995 and plucked him to safety.

The TRAP team went into Bosnia aboard two Sikorsky CH-53E Super Stallions, escorted by two Bell AH-1W Super Cobras and four McDonnell Douglas AV-8B Harrier IIs. On the way out, after picking up the downed pilot, the helicopters were engaged by one or more SA-7s and the door gunners returned fire.

The rescue was covered by two Aviano-based F/A-18D Hornets of squadron VMFA(AW)-533 'Hawks', which a week earlier had bombed a Serbian ammunition dump. Other Marines in the rescue were from the 24th Marine Expeditionary Unit (Special Operations Capable), Camp Lejune, North Carolina. The helicopters were part of HMM-263, a composite unit of UH-1N Twin Hueys, CH-53Es and AH-1Ws. The AH-1Ws in HMM-263 came from a detachment of HML-167 from MCAS New River, North Carolina. Also participating in the rescue were two A-10 'Warthogs' from an undisclosed squadron, an E-3 Sentry AWACS aircraft, EA-6Bs Prowlers from VAQ-141 'Shadowhawks' and VAQ-209 'Star Warriors', F/A-18Cs from VFA-15 'Valions' and VFA-87 'Golden Warriors'. Waiting in the wings, but not used, were EF-111 Ravens, F-15E Eagles and F-16 Fighting Falcons.

Open Skies aircraft

As part of preparations for entry into force of the 24 March 1992 Open Skies Treaty, which established a regime of unarmed observation flights over the entire territory of 27 signatory nations, the US and Germany conducted trial flights during 19-23 June 1995 with the first two aircraft modified to police the agreement.

The Boeing OC-135W (formerly OC-135B) 61-2674 of the US Air Force and Tupolev Tu-154M 11+02 of the German Luftwaffe are the first

two aircraft modified by signatory nations for Treaty enforcement. Russia is currently using an Antonov An-30 but is expected to shift to a Tu-154M based on the German modifications. Britain is using an Andover.

OC-135W 61-2674 gives the US initial operating capability for Open Skies. It is to be followed by OC-135W 61-2672 which was scheduled to complete modifications in September 1995 and to begin testing in January 1996. A third aircraft, 61-2670, will follow. When the latter two aircraft are in service, the US will consider that it has full operating capability and the first ship will be relegated to training.

The US is finding the OC-135W more expensive to operate than anticipated. A discussion has begun of replacing the OC-135W with a newer aircraft, probably the P-3C Orion.

T-3 Firefly mishap

The US Air Force's newest trainer, the Slingsby T-3 Firefly, was grounded for a week after a fatal crash on 22 February 1995 which killed an Air Force Academy cadet and a flight instructor. The mishap occurred at Colorado Springs, Colorado. The decision to ground all 30 aircraft in the service's T-3 fleet was made after the crash by the commander of the 12th Flying Training Wing Operations Group at Randolph AFB, Texas, which oversees T-3 operations. The grounding was described as a precaution stemming from the fact that the T-3 had been in USAF service less than a year.

T-3s are used to screen potential pilots by the academy's 557th Flight Training Squadron and by the 3rd Flying Training Squadron at Hondo Municipal Airport, Texas for Officer Training School and Reserve Officer Training Corps cadets. The USAF plans to buy 83 more T-3s. The USAF selected the Slingsby/Northrop T-67 Firefly as the replacement aircraft for its EFS (Enhanced Flight Screener) programme over several other candidates. Slingsby Aviation will build 113 Fireflys (down from an initial plan for

125), with a total programme cost of $54.8 million, for use in selecting pilot candidates at Hondo, Texas and the Air Force Academy at Colorado Springs. The aircraft will be assembled by Northrop at Hondo and delivered in 1993.

The Firefly is replacing an ageing fleet of 98 Cessna T-41A/C aircraft used to weed out unsuitable candidates for pilot training (T-41A) and at the Academy (T-41C). Screening of pilot candidates is aimed at reducing overall training costs.

Lockheed ASTOVL aircraft

On 1 May 1995, Lockheed Martin delivered its 86 per cent scale ASTOVL (Advanced Short Take-Off Vertical Landing) aircraft for the JAST (Joint Advanced Strike Technology) programme to NASA's Ames-Dryden Research Center, Edwards AFB, California. Initial hover tests began in July. The unmanned aircraft, one of several planned in a family of JAST prototypes, is powered by a 23,450-lb (104.31 kN) thrust Pratt & Whitney F100-PW-220 engine that provides both conventional thrust and shaft power to drive a lift fan. A full-scale version, if proceeded with, would be powered by the 35,000-lb (155.69-kN) thrust Pratt & Whitney F119 engine used on the F-22.

F-22 weight issues

Proponents of the Lockheed F-22 claim that weight increases in the evolution of the aircraft are 'fairly insignificant'. The fact that considerable attention has been paid to them, however, is a sign that they are not. The F-22 now has an empty weight 1,361 lb (617 kg) heavier than original projections. The F-22's wing was originally

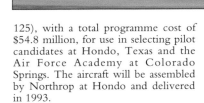

all-composite including skins, ribs and spars, with only the front and rear spar made of titanium. When a high-explosive 30-mm shell was fired into a wing section full of water, the test wing blew itself apart. As a result, structural pieces of the wing will now be surrounded with titanium to contain damage from a high-explosive shell. The current wing adds 129 lb (58.51 kg) to the weight of the F-22.

The General Accounting Office (GAO) says that the F-22 programme should be limited to production of six to eight aircraft per year in order to cut costs during flight testing and early operational testing. The USAF's plan is to manufacture 36 aircraft per year. GAO critics say the F-22's integrated avionics, super cruise propulsion and stealth technologies hold a high potential for costly developmental problems. According to production plans, 80 F-22s will be built before flight testing is completed.

V-22 construction

The Bell/Boeing V-22 tilt-rotor programme will include 48 aircraft for US Navy CSAR (Combat Search and Rescue), according to the US Marine Corps' V-22 programme manager. The 48 ships are in addition to 425 aircraft for the Marine Corps and US Air Force Special Operations Forces. Current plans are for procurement to last from 1997 to 2021. Marine Corps procurement will begin with four aircraft in Fiscal Year 1997. The first US Air Force CV-22 Osprey variants will be procured in FY 2001. Navy procurement will begin in FY 2009. Boeing has begun assembly of the forward fuselage of the first EMD (engineering and manufacturing development) V-22, which is the seventh machine to be manufactured.

UNPROFOR's own 'Air Force'

Quite apart from the NATO forces deployed for operations over Bosnia, there is another air force at work in the troubled state. The United Nations Protection Force has its own aircraft and helicopters, drawn from the air arms of several nations, and these provide constant support for the 'Blue Berets' on the ground.

Since 1992 UN military and civilian contract aircrews have been flying into some of the most dangerous 'hot' spots in the war-torn former Yugoslavia to keep the United Nations' peacekeepers and aid workers in business. When the United Nations Protection Force (UNPROFOR) was established in the spring of 1992, it was decided to rely on hired-in aircraft and crews as a means to cut costs. At the time, the UN was only policing the ceasefire line between Serbia and Croatia, and it was thought that civilian contractors would be able to cope with this 'routine' job.

Since then the troop strength of UNPROFOR has trebled and the mission area has grown to include all the republics of the former Yugoslavia. During 1994 alone, UN aircraft flew 20,500 flight hours, carrying 150,000 passengers, 250 medical evacuation patients and 25,000 tons of cargo, and undertook 1,200 missions to Sarajevo and 40 to Tuzla.

Early days

The first UN peacekeepers began to arrive in Yugoslavia during March 1992, but the notoriously slow UN bureaucracy meant it was the summer before the UNPROFOR's own 'air force' was fully operational. During this hiatus, UNPROFOR had to rely on the good will of the slowly disintegrating Federal Yugoslav Air Force for air transport, hardly an ideal situation.

Intercepted in the 'No-Fly Zone', this Croatian Mil Mi-8T 'Hip-C' was caught by an Army Air Corps Lynx over Gornji Vakuf. Despite its Red Cross markings, it was believed to be supporting Croatian forces inside Bosnia – in total defiance of the 'No-Fly' restrictions.

Left: The hard-working Army Air Corps Lynx AH.Mk 7s of No. 664 Sqn have swapped their regular home of Dishforth for a temporary detachment to Divulji Barracks, Split. The four aircraft active in Bosnia all carry an ALQ-144 infra-red jammer under their tail booms, and the standard exhaust shrouds around the engines. The Lynxes were airfreighted to Bosnia aboard RAF Hercules in January 1995.

Right: Four Sea King HC.Mk 4s of No. 845 Sqn, Royal Navy, are allocated to UNPROFOR in addition to those operating as part of Operation Sharp Guard aboard HMS Illustrious in the Adriatic. In addition to their regular ESM fit, the 'Junglie' Sea Kings have been fitted with chaff and flare dispensers (on the starboard side) and Loral ALQ-157 IR jammers behind the cockpit.

By the summer of 1992 contracts had been issued and UNPROFOR's 'air force' was in business, with Antonov An-26s and An-32s shuttling between Zagreb, Belgrade, Split and Sarajevo on passenger and cargo-carrying duties. Bell 212B and 206s were in service supporting UN troops in the UN Protected Areas of Croatia. Within months, the UN had hired giant Ilyushin Il-76 transports to supplement its 'air force' because of the need to supply its Sarajevo garrison by air. During late 1992 the first British and French military helicopters were deployed to Split to support the United Nation's Bosnia-Herzegovina Command.

Operations in 1995

In early 1995 the UN's air support consisted of six fixed-wing aircraft and more than 30 helicopters to move people and supplies around the former Yugoslavia, flying throughout Croatia, to Belgrade in Serbia, Skopje in Macedonia, and Sarajevo and Tuzla in Bosnia-Herzegovina.

The UN air operation is controlled from the G3 Air Branch in the Force Headquarters in Zagreb. Three air support groups are based at Pleso Airport, Zagreb, Skopje and Split. The number of tasks have grown to include logistic resupply of cargo and personnel, troop rotations, liaison, emergency casualty evacuations, airborne observation and patrol. Also based in Zagreb is the Military Close Air Support Co-ordination Centre, which links the UN with NATO's 5th Allied Tactical Air Force in Italy.

Below: UNPROFOR's first winter in Bosnia was as far back as 1992 and, for a few months each year since, the UN's white-painted helicopters have been a bit less of a target over the snow-covered hills of former-Yugoslavia. The arrival, in 1995, of the UK's 24th Air Mobile Brigade, as part of the United Nations' Rapid Reaction Force, means that British helicopters such as this Army Lynx AH.Mk 7 will be doubly busy.

The UN High Commissioner for Refugees' (UNHCR) airlift to Sarajevo (which sadly has been suspended for much of the latter part of 1995 due to fighting) is co-ordinated from the Air Operations Cell at UNHCR Headquarters in Geneva and at the Joint Air Operation Cell in Ancona, Italy, which controls western military airlifters taking part in the aid operation.

Worldwide support

ASG Pleso is the main hub for fixed-wing operations, as it is the base for chartered East European aircraft. The Ukrainian company Vitair provides two Il-76s, with a third on call, and a Tupolev Tu-154. Russia's Air Troika leases two Yakovlev Yak-40 and an Antonov An-26. Rotary-wing assets at Zagreb include two Sikorsky S-61s hired from Dutch airline KLM, two Air Troika Mil Mi-26s, two Bell 212s leased from the American firm Evergreen Helicopters Inc. and a Bell 206 hired from the Canadian company Skylink. Due to contractual reasons, the helicopters based at Pleso are only allowed to work within Croatia, mainly flying in support of UN troops in the Serb-occupied UN Protected Areas.

The UN's Former Yugoslav Republic of Macedonia Command is supported by ASG

Above: This ALAT Gazelle is one of the (unarmed) Athos scouts fitted with the SFOM sight above the cockpit. This sighting system is an integral part of the Gazelle's scout mission.

Skopje, which has two Skylink Bell 206s and an Evergreen Bell 212. Also based at Skopje are three US Army Sikorsky UH-60A Blackhawks of the 7th Battalion, 1st Aviation Regiment, but these remain under US control rather than working for the UN.

An Air Liaison Office is maintained in Belgrade to co-ordinate UN flights into the capital of Serbia. These are mainly Yak-40 flights carrying UN and international diplomats on peace missions.

Due to the greater threat over Bosnia, ASG Split controls the majority of the UN's military rotary-wing assets. G3 Air Ops in the Air Operations Co-ordination Centre (AOCC) at BHC (Forward) Headquarters in Sarajevo and at BHC Rear Headquarters in Split control air

Above right: This KLM Helicopters S-61N Sea King has abandoned its usual bright blue colours for the UN's white uniform. The 24-seat aircraft, which usually fly oil rig support missions, provide useful passenger transport and heavylift capacity.

Right: To the Norwegian armed forces, the Agusta-Bell AB 412 is known as the Arapaho. As an assault transport, the AB 412 can carry 14 troops and four aircraft are based at Tuzla in support of Norway's sizeable UNPROFOR force.

UNPROFOR's own 'Air Force'

Left: Ukrainian Il-76 freighters, such as this turret-equipped Il-76M, have played a substantial role in supporting UNPROFOR operations. This aircraft is seen unloading, behind a DHC-7 Dash Seven of Adria Airways, which is now the national airline of Slovenia.

Below: The Antonov An-12 'Cub' is one of the more senior types to wear the badge of the UN. This Russian example wears the digits of its (former Aeroflot) registration, but no national codes. The rear gun turret is a reflection of the An-12's military origins and is a common sight on many such commercially operated 'Cubs'.

operations over Bosnia. Divulji Barracks in Split is the main base for UN helicopters. The French Army Aviation (ALAT) 'Détachment Split' fields six Pumas and four Gazelles. Four British Royal Navy Sea King HC.Mk 4s of 845 Naval Air Squadron are also based at Divulji, alongside four Lynx AH.Mk 7s of 664 Squadron, 9 Regiment Army Air Corps. Based in Tuzla are four Bell 412 Arapahos which support the UN Nordic Battalion. To provide helicopter support for the Sector South West Headquarters, one British Sea King and two Lynxes are held on alert at Gornji Vakuf. A French Puma is held on alert at Kiseljak to support the UN garrison in Sarajevo. Four Royal Netherlands air force BO 105s of 229 Squadron are forward-based at Lukovac, near Tuzla, under Dutch national command.

Left: Antonov's An-26 tactical transport is well suited to operations in regions such as Bosnia. Airline-operated freighters, wearing United Nations' marks, operate alongside An-32s of the Ukrainian air force. The An-32 is a higher-powered version fitted with AI-20DM turboprops.

The UN air support operation is completed by a Movement Norwegian Control operation based at Pleso, with outposts in Sarajevo, Split, Belgrade, Skopje and Sarajevo to co-ordinate passenger and cargo manifests for all UN air operations.

During late spring 1995, as fighting increased in Bosnia and a reduction in peacekeeping troops in Croatia began, the UN air operations staffs were increasingly having to consider the logistics of moving almost 40,000 peacekeepers from the former Yugoslavia. This could be their greatest test. **Tim Ripley**

Top: This Russian Mil Mi-8 boasts a United Nations serial. Military Mi-8s can also be found in the hands of several of the warring parties in Bosnia.

Right: This Tupolev Tu-154M has been leased from Bulgaria's Balkan Airlines – a good way for the company to make money from the uneconomic former Soviet airliner.

Below: True heavylift capability is provided by the giant Mil Mi-26 'Halos' of Russia's Air Troika. These aircraft are 'home-based' at Zagreb and can conceivably carry up to 100 people.

BRIEFING

Lockheed Martin F-16

Reconnaissance Falcons

In September 1995 the US Air Force is due to retire its final squadron of RF-4C Phantoms (192nd RS/ Nevada ANG), and with it will go the USAF's sole manned tactical reconnaissance capability unless a replacement can be found. Faced with this unfavourable state of affairs, the USAF is now seeking a low-cost successor to the RF-4C that can be developed rapidly.

McDonnell Douglas has test-flown an F-15 Eagle with a belly-mounted pod containing sensors from the ATARS (advanced tactical airborne reconnaissance system) programme, but this represents a high-cost approach to the RF-4 replacement problem. Conversely, Lockheed Martin has fitted its F-16C with a much cheaper alternative and this has undergone operational trials with the Air National Guard.

Tactical reconnaissance has been a role associated with the F-16 for several years. Danish aircraft (with Red Baron pods) and Dutch machines (with Orpheus pods) currently fly the mission, while in the early 1980s General Dynamics developed a dedicated RF-16D testbed with a conformal belly pod housing a multi-sensor suite based on ATARS. This featured (from front to back) low-altitude and medium-altitude sensor bays, infra-red linescan and associated systems for environmental control and data recording.

Lockheed's current reconnaissance programme features the same pod as was carried by the RF-16D testbed, but without the linescan and datalink functions. Utilising the

Right: This RDAF F-16 carries a mock-up of the Per Udsen MRP, which is due to make its first flight in early 1996. The pod can be configured for a wide variety of sensors.

same pod has alleviated the need for aerodynamic and structural clearance work. Inside the pod is the standard KS-87 camera from the RF-4C, but with an EO video back replacing the wet film cartridge used previously. The pod can be attached to the centreline of any F-16, and will be capable of carrying far more advanced sensors in the near future, such as the ATARS suite (consisting of a low- and medium-altitude EO imaging system and an infra-red linescan) or the Loral AVD-5 EO-LOROPS camera (as fitted in the nose of some RF-4Cs) should these form part of a future requirement.

The programme is a team effort. Lockheed provides the pod, aircraft modifications (which include the provision of GPS and a control panel) and overall systems integration. The US Air Force provided the aircraft and the KS-87 camera, while other suppliers are Recon/Optical Inc. (electro-optical video back), Ampex Corporation (DCRsi-240R digital data recorder) and TERMA Elktronik (cockpit controller).

Tests have been conducted by the 192nd FS/Virginia ANG with one F-16D, which first flew with the pod on 26 April 1995. The aircraft has achieved outstanding results, fully validating the low-cost, low-risk approach. The operational trial is expected to last until July.

Lockheed Martin is not the only company involved in F-16 recon-

naissance options. The Danish component manufacturer Per Udsen (which has also developed an F-16 pylon with integral chaff/flare dispensers) has developed a similar Modular Reconnaissance Pod (MRP) for the F-16. This is based on a different structural philosophy where the sensors are hung from a strong upper beam rather than being supported on the floor of the pod. This allows easy access to the sensors and a lighter pod weight.

The Royal Danish air force has procured three of these pods for evaluation on F-16s, the sensors being Vinten wet-film cameras

El-Op technicians are seen with the company's LOROP camera and a mock-up of the modified centreline fuel tank which contains it. The El-Op sensor can work in conjunction with a datalink to transfer imagery to ground stations.

removed from the current RDAF Red Baron camera pod (see *World Air Power Journal* Volume 20 p.4). First flight for a reconnaissance-configured F-16 is slated for early 1996, with a production decision expected in 1997. The Per Udsen pod is also to be evaluated for the USAF by the ANG/AFRes Test Center, part of the Arizona ANG organisation at Tucson. Flight trials are to begin in early 1996 with sensors supplied by CAI. Like the Fort Worth pod, the Per Udsen unit is compatible with a wide range of sensors. The Danish company is currently looking at the Loral ATARS suite and also at the Israeli El-Op LOROP camera. This sensor, similar to the AVD-5, is also believed to be carried by IDF/AF F-16s in a modified centreline fuel tank.　　**David Donald**

Airborne testing of Lockheed Martin's low-cost pod has been entrusted to the Virginia ANG with this single F-16D. With the need to replace RF-4s looming in 1995, Lockheed contends that the main issue is to provide a cheap yet effective reconnaissance capability which is available in the very short term. This pod, with modified existing cameras, fills this requirement admirably, while still offering the option for more sophisticated sensors in the future.

Lockheed Martin/Boeing Tier 3-
DarkStar UAV unveiled

Previously known as the Lockheed Advanced Developments Company, the Lockheed Martin Skunk Works has a reputation second to none for producing weird and wonderful aircraft. The latest offering to be made public is the DarkStar unmanned air vehicle, produced in partnership with Boeing. This UAV is expected to greatly enhance the high-altitude surveillance capability of US forces, being able to operate in hostile airspace denied to conventional assets. It forms part of a two-pronged programme being undertaken jointly by the DARO (Defense Airborne Reconnaissance Office) and ARPA (Advanced Research Projects Agency). The DarkStar was unveiled on 1 June 1995 at the Skunk Works' Palmdale facility.

Partnering the DarkStar is the Teledyne Ryan Tier 2+ UAV, which is a non-stealthy long-endurance platform with high altitude capability (24-hour endurance at 65,000 ft/19812 m). The Ryan machine is intended for surveillance in low threat density areas, while the Tier 3- is optimised for shorter missions into highly dangerous areas when the need arises. Consequently it has been designed with low observable characteristics as the main design driver.

The ideal shape for a loitering air vehicle in terms of stealth would be a flying saucer, and to that end the front of the DarkStar's fuselage body is semi-circular in planform. This gives a 'fuzz-ball' return on radar screens, which has been further reduced by the considerable depth of RAM incorporated into the fuselage body design. The high-aspect ratio wings have straight edges which produce two strong radar spikes fore and aft of the aircraft, but by careful mission planning these spikes can be directed away from any threat radars.

Design of the DarkStar is directed primarily at avoiding attack by fighters, which are seen as the UAV's greatest threat. Consequently the aircraft is most stealthy in the UHF band used by early warning radars, and the X-band used by fighter radars.

The fuselage body contains the undercarriage, sensor, communications and engine. The intake for the single Williams FJ44 turbofan is in the fuselage upper surface, and the powerplant exhausts through a rectangular slit in the rear. Either side of the intake are two small sta-

tic ports to provide air data for the flight control system. Despite its unusual appearance, the DarkStar is aerodynamically stable, the wings having a slight forward sweep keeping them near the centre of gravity while a reflexed trailing edge provides a nose-up pitching moment offseting the mass of the nose.

Much of the design was derived from technology developed over more than a decade of research by both companies, notably Boeing which had earlier produced the Compass Cope and Condor UAVs. The design shape was frozen just 11 days after ARPA issued the contract on 20 June 1994. An aerodynamic problem which surfaced in late 1994 was solved by the insertion of an additional 2-ft (0.6-m) section in each wingroot, raising overall span to 69 ft (21 m). Control surfaces consist of two elevons inboard on each wing, and an outboard B-2-style split-flap ruddervon. For take-off, the nosewheel strut is extended to raise the wing incidence. This allows a rotationless take-off, as the elevons do not have sufficient authority to unstick the aircraft from a zero-incidence condition.

At the crown of the humped fuselage is a flat antenna for a Ku-band satellite communications antenna, allowing imagery to be transmitted in near real time to ground stations. A UHF satellite antenna and GPS receiver are located behind the main antenna. In addition to satcoms there is a wide-band line-of-sight downlink to ground stations.

Two prinicpal sensors are envis-

aged for the DarkStar: the Westinghouse Synthetic Aperture Radar (as originally developed for the cancelled A-12 programme) and the ReconOptical electro-optical sensor. Coverage is in the region of 1,600 nm² per hour with a 3-ft (1-m) resolution, with the capability to perform 600 high-resolution (1-ft/0.3-m) 'spots' in a single mission. Both sensors will peer to port, and the sensor pallet is easily accessible via a panel on the left side of the fuselage. The avionics pallet is accommodated to starboard.

Having only just emerged from the 'black' world, there are few exact details concerning the DarkStar's performance. Of the 8,600-lb (3900-kg) maximum take-off weight, about 3,000 lb (1360 kg) is fuel, giving an endurance of over eight hours at a 500-nm (575-mile; 926-km) radius with the SAR sensor, which together with the communications equipment weighs 1,000 lb (454 kg). The EO sensor is some 200 lb (91 kg) lighter, resulting in an endurance increase of about 90 minutes.

Service ceiling is over 45,000 ft (13716 m), placing DarkStar out of the envelopes of most surface-to-air missiles. However, it is still in the subsonic fighter intercept envelope, and the Lockheed Martin/Boeing team have expressed a desire for the aircraft to be given a

"A flying saucer with wings attached" – seen during its 1 June public roll-out at Palmdale, the DarkStar displays its futuristic configuration.

65,000-ft (19812-m) ceiling – like its Tier 2+ partner – to place it into the supersonic engagement regime, where the chance of a fighter kill is far less. The wings are detachable for rapid air transportability. The upper surfaces of the DarkStar are white to avoid overheating from the sun, while the underside is black for low visibility from the ground.

Around 20 of each of the Tier 2+/Tier 3- drones are required, with a fixed unit flyaway cost of $10 million. The company workshare is 50:50, Boeing manufacturing the wings while Lockheed is responsible for the body. The first of two prototypes is to be shipped to NASA Dryden at Edwards AFB, for a first flight scheduled for September 1995. In the spring of 1996 the No. 2 prototype will undergo RCS testing at Helendale, California. The programme is currently ahead of schedule, and service entry is expected by 1999 at the latest, some officials hoping it will be 1997.   **David Donald**

The prototype Tier 3- carries virtually no markings, apart from the serial '695' and a small skunk motif on the dorsal blade aerial.

Northrop Grumman/Boeing E-8

J-STARS nears service

Northrop Grumman's Electronics and Systems Division at Melbourne, Florida, is continuing to develop the E-8 J-STARS system prior to its full operational debut expected in 1997. Three E-8 airframes are currently flying: the two original E-8As and the first production-configured E-8C. Funding presently covers eight operational aircraft, although the initial requirement is for 20 aircraft plus the first E-8C as a permanent test platform. The two E-8As will be modified to operational configuration later in the production schedule. Unlike most other US defence programmes, J-STARS has drawn little opposition from politicians, and is likely to receive full funding for the first batch, with a follow-on order a probability.

NATO is keen to acquire a battlefield surveillance system, and the J-STARS is a candidate for this requirement. The choice of the E-8 would be logical as the aircraft could share maintenance facilities with the NAEWF which operates the 707-based E-3 Sentry. Following a successful European theatre evaluation in 1994, an E-8A was displayed at the 1995 Paris air show to further press the J-STARS case

The cockpit of the E-8A has changed little from its Boeing 707 origins, although military systems have been added. In the development aircraft the navigator still sits on the flight deck, although for operational aircraft the station will be moved back into the main cabin.

for the NATO requirement.

Externally the aircraft has changed in one significant respect from the early configuration used in Desert Storm (described in *World Air Power Journal* Volume 9). The large underbelly fairing for the 'Fiddle' Flight Test Datalink antenna has been removed. The tiny 'Skittle' Surveillance and Control Datalink antenna remains under the rear fuselage, this being the primary antenna for transmitting data to the GSM (Ground Station Module).

When the system becomes operational, GSMs will be issued to ground formations down to brigade level, and also to specialised units such as Military Intelligence Battalions. The GSM receives data not only from the J-STARS, but also from UAVs, Sigint platforms such as the RC-135 Rivet Joint and RC-12 Guardrail, and Trojan Spirit satellite intelligence. GSMs can be linked by cable to UAV operating stations so that the efforts of the drones and the J-STARS can be coordinated. At command level, the GSM vehicle module can be positioned up to 1,000 ft (305 m) away from the command console, allowing intelligence displays and system commands to be conducted in operations rooms rather than at the GSM vehicle.

Naturally the E-8As are still largely full of test equipment, but

some production standard operator consoles are now flying. Additionally there are highly advanced experimental consoles which are paving the way for ongoing post-IOC developments. The consoles provide monochrome images from the SAR function of the radar, over which can be laid the colour display from the MTI function. The system has the ability to display small windows on the screen with displays of different scale.

For the production E-8C, the navigator function will be moved out of the cockpit and into the main cabin, where there are 17 operator consoles. Precise positional information is a prerequisite for the J-STARS system to work accurately, and the relocation of the navigator allows him to work closely with the system operators. A dedicated communications console will be provided for the dissemination of information, including via the secure Link 16 JTIDS. Incoming data, such as system commands from the GSMs or situation data from an E-3 AWACS, is also processed through this console. The production E-8C will have defensive systems, but precisely which has yet to be defined.

While the joint Northrop Grumman-US Air Force test team continues its development work, the two E-8As are still ready to deploy at very short notice should they be needed. IOC is expected in 1997 with five airframes. Robins AFB, Georgia, will be the operational base, and the unit is expected to be the 6th RW. **David Donald**

Above: This console is displaying a typical synthetic aperture radar patch map, depicting a bridge over a major river. The moving target returns could be overlaid to show the traffic on the bridge if required, while the screen could use a sub-window to display an image of a different scale. A history replay function allows the operator to review the display in quick time to spot trends that may not be obvious in real time.

Below: This is the dedicated communications console, displaying the available radio/datalink equipment and the 16 radar operator stations. The aircraft has 16 UHF channels, two HF and three VHF, in addition to JTIDS.

One of the two E-8A development aircraft (named 'Night Stalker') was on display at the 1995 Paris air show. The main J-STARS radar array is in the canoe fairing under the forward fuselage. The aircraft flies an elongated racetrack pattern, with the radar looking sideways. As the aircraft completes a 180° turn at the end of the racetrack, the antennas switch their attention to the opposite side of the aircraft to maintain coverage of the original area of interest.

Mirage 5M Elkan

Elkan supersedes Hawker Hunters

Further to the feature on Chile's new Elkans in *World Air Power Journal* Volume 22, Grupo 8 of the Fuerza Aérea de Chile has taken delivery of its first fully modified aircraft. Although several aircraft flew in Belgium in full Chilean markings, some of them used by Détachment MirSIP, the conversion training unit, these were Belgian-standard MirSIPs and not fully converted to full Elkan standards.

The aircraft formally handed over in Chile were full-standard Elkans, two-seater 716 (former Mirage 5BD BD-01) and single-seater 703 (former Mirage 5 BA BA-11). The only external difference between the MirSIP and the Elkan is the VOR/DME antenna on each side of the fin, and the lack of Belgian serials and fin flash.

Delivery of the Elkans allowed Chile to retire the last of its ageing Hunters, leaving Zimbabwe as the only remaining front-line operator of the classic Hawker fighter. Chile's Hunters were given a cere-monial retirement, with an official last flight in the hands of some of the country's most senior military aviators. The four aircraft (two twin-stickers and a pair of single-seaters) were flown by a gathering of top brass, all of whom had flown the Hunter earlier in their illustrious careers. The Commanding Officer

of Grupo 8, Comandante de Grupo Patricio Saavedra was joined by the air force commander-in-chief, General del Aire Ramón Vega Hidalgo, the Chief of the Air Staff, General de Aviación Fernando Rojas Vender, and Generals de Aviación Jaime Estay Viveros and Enrique Montealegre and General de Brigada Florencio Dublé. In fact, although this was officially the Hunter's last flight in Chilean ser-vice, the aircraft may remain in use for some months while the Elkans take over, and at least two Hunters will remain airworthy in Chile, and will be used as hacks as well as his-toric display aircraft. Interestingly, the two single-seat Hunters which took part in the final flypast in April

were both fitted with camera noses. Until Chile receives the four unconverted Mirage 5BRs these will be Chile's only tactical recon-naissance aircraft, and may thus be presumed to be most likely to be the last of the Grupo 8 Hunters to be retired.

Jon Lake

Right: The Elkan badge applied to the tailfins and noses of Chile's new Mirages. The aircraft is officially known as the Mirage 5M. The name Elkan means guardian in a local Indian dialect.

Below: Surrounded by Hunters from Grupo 8 and F-5Es from the co-located Grupo 7, the first two Chilean Elkans are formally accepted by Grupo 8 at Basa Aérea de Cerro Moreno, Antafagasta.

Above: One of Chile's last Hunters, a single-seater with a recce nose. Chile's initial batches of Hunters totalled 39 aircraft (mostly FGA.Mk 71s, but with six FR.Mk 71As and four T.Mk 72s), and these were augmented by 12 ex-RAF FGA.Mk 9s delivered after the Falklands War.

The RAF's support helicopter force has largely escaped the cuts which have beset the fast jet community, since the air mobility it provides for the Army has become more important in the post–Cold War world. It has been a closely guarded secret, but the support helicopter world gives pilots and aircrew access to almost limitless low flying, and imposes an incredible requirement for initiative.

Above left: RAF Odiham is the home of the Royal Air Force's battlefield support helicopter force. Resident units include No. 7 Squadron with Chinooks, No. 33 with its Puma HC.Mk 1s and No. 27(R) which trains aircrew for both types. Two of No. 7 Sqn's Chinooks are detached to Aldergrove, as part of the support helicopter Force Northern Ireland. The Chinook HC.Mk 1 (CH-47C) fleet is being upgraded to HC.Mk 2 (CH-47D) standard and most have now been redelivered by Boeing.

Left: Puma crewmen are all trained to use the 7.62-mm GPMG machine-gun (or 'gimpy'). This weapon was regarded as being especially important in Northern Ireland, where the danger of coming under hostile fire was ever-present.

RAF Support Helicopters in Action

Above: A Puma HC.Mk 1 of No. 240 OCU closes to within a one-rotorspan separation with a sister ship, to recover to its Odiham base. Forty Pumas were delivered to the RAF, beginning in 1970. The type's twin doors and spacious cabin make the type popular with its army users, who also appreciate the versatility. Today, four squadrons are wholly or partially equipped with the type: Nos 18 (RAF Laarbruch), 27(R) (RAF Odiham), 33 Sqn (RAF Odiham), and 230 (RAF Aldergrove). In 1994 the Pumas of No. 1563 flight returned from Belize International Airport to the UK, and the Laarbruch-based aircraft will also return in the near future. RAF Pumas have undergone a host of modifications and improvements during their long career.

Above and right: A Puma dispenses IRCM (infra-red counter-measures) decoy flares over a range in the English Channel. The Puma's defensive suite was considerably upgraded during the 1980s, not least for its participation in operations in Northern Ireland, with chaff/flare dispensers, missile warning equipment and radar warning receivers. These improvements were adopted fleet-wide, however, and dramatically reduced the aircraft's vulnerability to hostile groundfire. Aircrew report that you can feel the heat these flares generate from inside the cabin.

Above left: A Wessex HC.Mk 2 of No. 2 FTS at Shawbury hovers during underslung load training. Helicopter pilots transition to the Wessex from the Gazelle, learning operational flying techniques on the larger, older type. The Wessex also remains in front-line service with squadrons in Cyprus and Hong Kong, and with two UK-based support helicopter units, and with one flight of No. 22 Squadron in the search and rescue role.

Left: A student crewman voice marshalls the pilot of a No. 2 FTS Wessex over the intended landing point for the underslung trailer. The ageing Wessex has some demanding handling characteristics which make it an ideal trainer.

Above: A Puma HC.Mk 1 and a Chinook HC.Mk 1, both from the Odiham-based No. 240 OCU, undertake a mixed formation exercise. The unit, which has been redesignated as No. 27 (Reserve) Squadron, trains pilots and aircrew for both types in entirely separate courses.

Right: The RAF's Wessexes and Pumas began receiving a new two-tone green camouflage during the early 1990s, in a programme which is continuing, but still incomplete. The Puma has lost its black belly. Here a Puma of No. 240 OCU shows off the new colour scheme.

Above: No. 18 Squadron was the first of the RAF's Chinook squadrons to form, on 4 August 1981, and initially operated from Gütersloh in Germany. With the contraction of RAF Germany the squadron picked up Pumas from No. 230 Squadron, and moved to Laarbruch. The front door step is often used by the No. 2 crewman to help with voice marshalling.

Left: A Chinook of No. 240 OCU demonstrates that you don't have to fly fighters to take advantage of aerodynamic braking.

Right: A pilot from No. 7 Squadron perches the front wheels of his Chinook on an outcrop in the Falklands during late 1982. During the Falklands War earlier that year, a single Chinook performed epic feats of endurance.

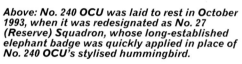

Above: No. 240 OCU was laid to rest in October 1993, when it was redesignated as No. 27 (Reserve) Squadron, whose long-established elephant badge was quickly applied in place of No. 240 OCU's stylised hummingbird.

Above right: This No. 230 Squadron aircraft went to war in the Middle East, albeit with a coat of desert pink over its anniversary colour scheme. For many years No. 230 was the RAF's only Tiger squadron.

Right: Anyone for croquet? A Puma demonstrates that even a lawn can serve as an airstrip for the support helicopter force, and that a Puma can be a suitable conveyance for lunch in the country – here, at a country house in Devon.

Above left: A No. 72 Sqn Wessex backs away from its parking slot, while a Royal Marines Gazelle hovers in the background. Reports suggest that surplus Army Lynxes could replace the Wessex.

Left: This unique No. 33 Squadron Puma is configured for VIP duties, and has enlarged Super Puma-type sponsons housing extra fuel to extend the aircraft's range.

Above: A Chinook HC.Mk 2 carries netted underslung loads. The HC.Mk 2 has a maximum all-up weight of 22,700 kg, making it a truly impressive lifter.

Right: Dusty conditions and rotor downwash create a 'brown-out' which affords the deplaning soldiers some limited degree of cover. The RAF's Pumas saw active service in the Gulf War, where they performed valiantly despite their age and the type's perceived fragility.

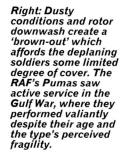

Left: Three Pumas of No. 230 Squadron spiral into a clearing for an assault landing, during an exercise with No. 24 Airmobile Brigade exercise. For many years No. 230 Squadron provided helicopter support for the BAOR (British Army Of the Rhine) from its base at Gütersloh. It has since moved back to the UK and is presently based at RAF Aldergrove in Northern Ireland. The RAF has a complex concept for using helicopters, operating them to move maximum resources in minimum elapsed time, while exposing them to minimum danger.

Left: The Sea King HAR.Mk 3s of No. 78 Squadron in the Falklands fly both search and rescue and support helicopter missions. Unlike UK-based Sea Kings which are painted in high-conspicuity yellow, they wear an overall sea grey colour scheme, with toned-down markings. No. 78 Squadron operates a mixed fleet of Sea Kings and Chinooks.

Right: A No. 78 Squadron Chinook HC.Mk 1 drops off a Haglund BV206 Snowcat at its Mount Pleasant base, against a threateningly stormy South Atlantic sky. The squadron has now swapped its Chinook HC.Mk 1s for upgraded HC.Mk 2s.

Above: This Chinook is fitted with a streamlined rescue winch above the forward door, and has a Nitesun searchlight in an articulated mounting below the nose. Chinooks are equipped with a sophisticated AFCS and always carry out flying control checks on take-off. This aircraft wears the badge of No. 78 Squadron on its nose.

Right: A Chinook HC.Mk 1 of No. 1310 Flight does underslung load work aboard the commercial container ship MV Corato during 1984. No. 1310 Flight was based at San Carlos, and formed the basis of No. 78 Squadron. There has been a permanent Chinook presence on the Falklands since the end of the war. The Chinook is fitted with a versatile triple hook system, giving great flexibility. The crewman can be seen craning his head out of the forward door, it being difficult to see underslung loads.

Left: Still wearing Gulf War-type special night camouflage of desert pink mottled with black, this RAF Chinook HC.Mk 1 is fitted with 'drainpipes' to catch and divert spent shell cases from the door-mounted machine-guns. The aircraft is also fitted with an IR jammer and sundry chaff dispensers. This unusual aircraft was pictured operating near Silopi in Turkey, during humanitarian relief operations in support of Operation Haven. The exact role played by similarly camouflaged and similarly equipped Chinooks in the Gulf War has still not been officially revealed.

Left: No. 240 OCU provided the first three Chinooks sent out to Southern Turkey/Northern Iraq for Operation Haven. This Union Jack-bedecked machine is seen 9,000 ft up somewhere in Iraq. Prominent Union Jacks were earlier applied to RAF Chinooks for operations over the Lebanon during the mid-1980s. The RAF's Chinooks are currently completing an upgrade programme which will bring them up to HC.Mk 2 standards, with a new AFCS, upgraded hydraulics, engines and transmission, and with airframe, avionics and rotor improvements. The HC.Mk 2s are redelivered in a low IR signature overall dark green colour scheme. The HC.Mk 2 is broadly equivalent to the definitive US Army CH-47D standard.

Above: Some 5000 kg of building stores are seen en route to a refugee camp in Iraq, courtesy of a No. 240 OCU Chinook HC.Mk 1. The Chinook's rugged dependability, capacious cabin and awesome lifting capabilities have endeared it to its RAF operators and to its Army users. Despite the age of the basic Chinook design, the aircraft's latest variants remain unmatched in many respects. Thirty-two Chinooks are being upgraded to HC.Mk 2 standards, and nine more new-build HC.Mk 2s have since been ordered, along with eight HC.Mk 3s, which will be equivalent to the Special Operations-configured MH-47E. It is not known whether the latter aircraft will be fitted with refuelling probes.

Right: A Puma of the Odiham-based No. 33 Squadron brings in a 105-mm Pack Howitzer during a JATE (Joint Air Transport Establishment) exercise. Pumas also serve with No. 230 Squadron in Northern Ireland and with No. 27 (Reserve) Squadron. The RAF's next-generation support helicopter will be the three-engined EHI EH101, which is expected to enter service with No. 60 Squadron at RAF Benson. Procurement of the tried-and-trusted Boeing Vertol Chinook is still continuing.

Saab 105/Sk 60 Variant Briefing

Little-known outside its native Sweden, and almost unheard of beyond Europe, the Sk 60 is one of Saab's most enduring designs. From its roots as a jet trainer, the Saab 105 has spawned attack, reconnaissance, even transport versions, and with the Swedish air force it will be training Gripen pilots into the next century.

Sweden has been at the forefront of aerospace development in Europe since the end of World War II. The swept-wing J 29 Tunnen and the transonic all-weather J 32 Lansen were remarkable achievements, while the J 35 Draken was one of Europe's earliest Mach 2 fighters. The de Havilland Vampire Mk 50 (J 28B) and T.55 (Sk 28C), along with the piston-engined Saab Sk 50 Safir, were then just adequate for pilot training but the introduction of the Draken, and of the planned J 37 Viggen, would demand a more modern training aircraft.

In the late 1950s, Saab had started studying a small business jet design, in line with similar developments in the USA. Saab drew up plans for a radical-looking, Delta-winged aircraft with canard foreplanes, seating five passengers. They kept an eye on the air force's need for a trainer and hedged their bets by investigating both the civil and military applications of the design. Saab's expertise lay primarily in the military field and it had enjoyed little success with civil types. As a result, in April 1960 Saab relaunched a redesigned (more conventional)

A pair of Austrian Saab 105Ös pushes over the top of a roll. Austrian aircraft are fitted with more powerful General Electric J85 engines, which provide extra margin for high-altitude operations in Austria's mountains.

Saab 105 trainer as a private venture, with Ragnar Härdmark as project manager.

Almost simultaneously the Swedish air force announced that it was seeking a new jet trainer, to take students through to the Draken. Changes were made to accommodate a vaguely stated requirement for a secondary light attack capability. The original design's 1,065-lb (4.7-kN) Turboméca Marboré 6 turbojets were replaced with more powerful, less thirsty Aubisque turbofans. The Aubisque (Swedish designation RM9) was derived from the 'hot section' of the French manufacturer's Bastan turboprop and was developed with funding from the French and Swedish governments. The General Electric J85 was considered but ruled out on cost grounds. Several rivals were evaluated by the Swedish air force (Flygvapnet), including the Hunting Jet Provost, Fouga Magister, Macchi M.B.326 and Canadair Tutor. Despite existing only on paper, the Saab 105 was selected and the Swedish government authorised production of the first prototype on 16 December 1961, signing a letter of intent for the purchase of 100+ production aircraft.

A conventional shape

The Saab 105 was a high-winged, T-tailed aircraft, built around an all-metal, semi-mono-coque fuselage. The short-stroke tricycle undercarriage retracted into a squat fuselage carrying two occupants on side-by-side seats and with two integral fuel tanks, further back. Two additional fuel tanks were located in the wing. The engines were shoulder-mounted, and fitted with idling thrust spoilers. Perforated airbrakes, in pivoting transverse slots, situated on the lower fuselage, behind the landing gear. Almost 10,000 man hours were spent in ground engineering tests before the aircraft flew. A static test airframe was used to validate design loads. At maximum design load, the relatively short wing deflected up to 18.6 cm (7⅓ in), but it did not fail until loads reached 163 per cent, with the wing then bent by 31 cm (1 ft ½ in). Drop tests cleared the landing gear to a sink speed of 3.7 m/s (12 ft/s) at 163 kt (230 km/h; 143 mph). On a rotating test rig, the fuel system was subjected to temperatures varying from -40° C to +40°C at simulated altitudes reaching up to 12000 m (39,370 ft). High-speed descents from 12000 m to 3000 m (39,370 ft

to 9,842 ft), and prolonged inverted flight were also simulated. The cockpit was subjected to 45,000 trial pressurisations, without failure. Eleven live firing ejection tests were made with and without the canopy and the Saab-built seats were cleared down to heights of 70 m (230 ft).

Several improved versions

Saab soon began developing further versions for the export market. The 'biz jet' concept was resurrected briefly with the Saab 105C proposal, which never got beyond the drawing board, but Saab made rapid progress in developing a re-engined multi-role version. The Saab 105XT was powered by a pair of General Electric J85 turbofans and was ultimately sold to Austria. This prompted further developments, which culminated in the Saab 105G light attack aircraft, but Austria remained Saab's only export 105 customer. Sweden's tradition of neutrality has always hampered Saab's attempts at marketing its aircraft to foreign customers – in effect, the company was barred from selling combat aircraft to nations that might actually need to use them. The 105XT's export chances were further hampered by potential restrictions on its American-supplied engine, much like the JAS 39 today.

In Flygvapnet service the Saab 105 was to have been replaced by the Saab B3LA. This was intended to serve as an attack aircraft, supplanting the AJ 37 Viggen, and to have a secondary training role. The B3LA was designed around a super-critical wing and a FLIR system for night/bad-weather operations, and it would have been highly manoeuvrable at low level, while possessing acceptable handling characteristics for the training role. The type was cancelled in 1979, leaving the Sk 60 to soldier on.

By the late 1980s major work was needed on Sweden's surviving Sk 60s. Between 1988 and 1991, a programme of structural modifications (including rewinging) was therefore initiated. This significantly increased the fatigue life of the 142 aircraft involved. The ejection seats were not replaced, but were modified to allow a new parachute to be accommodated. This structural overhaul came as a precursor to the complete re-engining of the surviving Sk 60s with the Williams/Rolls-Royce FJ44 turbofan. This was announced at the 1993 Farnborough air show and by the latter half of

1995 the first aircraft had begun its test programme with the new powerplants. Work is being carried out by the air force, at its Ljungbyhed workshops, in conjunction with Saab at Linköping. This move will dramatically improve the Sk 60's performance, while reducing running costs and noise output.

The Saab 105's last export opportunity came in 1986/87, with the launch of the USAF PATS primary trainer competition. PATS followed the New Generation Trainer programme, which had spawned the Fairchild T-46. The Saab 105 was also a side-by-side design, so for PATS Saab proposed a re-engined version of the 105, the Saab 2060, with new avionics and a new escape system. When the US Navy joined PATs in 1990, (resulting in JPATS), a Joint Statement of Operational Need was issued which specified 'stepped tandem seating' and a pressurised cockpit, ruling out the Saab proposal.

The Sk 60 continues in mainstream service with the Flygvapnet, and the FJ44-powered Sk 60(W)s will serve until 2010 or 2015. The service introduction of the Gripen in 1996/97 will reduce the usefulness of the Sk 60, which cannot provide an adequate preparation for pilots destined for the CRT-equipped front-line fighter. It is therefore likely that the Sk 60 will receive a substantial avionics upgrade. The air force is even investigating expanding the roles of the aircraft. While the army has launched an evaluation of the AH-64A Apache and Mi-28 'Havoc' attack helicopters, to replace existing TOW-equipped MBB 105s, the Flygvapnet has launched a study into using the Sk 60 as a dedicated CAS aircraft, to fulfill the same requirement.

Robert Hewson

Saab 105 prototypes

On 29 June 1963 Karl-Erik Fernberg flew the first Saab 105 (105-1/SE-501) at Linköping. The accelerated 120-hour flight test programme continued until 15 February 1964, highlighting several significant design problems. The wingroot and engine inlet required a substantial redesign and SE-501 was grounded for modification. The second prototype (105-2/SE-502), in the revised configuration, first flew on 17 June 1964. The two prototypes accumulated 1,000 hours flying time in nearly 1,300 flights. SE-501 became SE-XBY in March 1965, but was lost during inverted spinning trials, on 17 June 1966, at Ekenäs Gård. Eric Sjöberg ejected and later went on to become the first Saab 2000 pilot.

SE-502 was reregistered SE-XBZ and attended the Paris air show in 1971 (show number 751), 1973 (416) and 1975 (342). In 1972 this aircraft attended the Farnborough show and was forced to make an emergency wheels up landing, during the event. The Saab 105 survived to return to Farnborough in 1974 and 1976. It then went on to serve as the Saab 105XT and Saab 105G prototype. SE-XBZ made its last flight, a short hop from Linköping to the air force museum, on 29 October 1992.

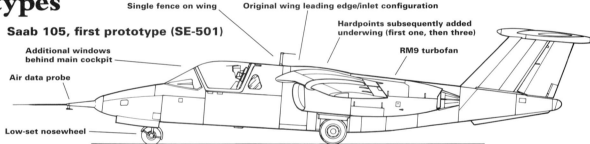

Saab 105, first prototype (SE-501)

Single fence on wing
Original wing leading edge/inlet configuration
Hardpoints subsequently added underwing (first one, then three)
Additional windows behind main cockpit
RM9 turbofan
Air data probe
Low-set nosewheel

Below left: SE-501 inherited its basic layout from its origins as an executive jet, but this soon proved to be less than desirable. Initial flight test experience forced a rapid redesign.

Below: Right from the outset, SE-502 flew with entirely redesigned inlets, jet pipes and engine cooling scoops along with a revised wing, featuring two fences. This was the definitive Sk 60 layout.

Saab 105C/D

At the 1966 Hannover air show Saab exhibited a model of the Saab 105C, a four- to five-seat business jet. The large canopy of the military trainer was replaced by a faired-in cabin roof, allowing an air-conditioned, pressurised fuselage. A door, with an integral air stair, separated the front and rear seat passengers, who were accommodated on a bench-type seat and provided with a window above the engine intakes. The wingspan was increased to 11 m (35 ft 9 in) to allow extra fuel tanks to be fitted. The Saab 105C was pegged at £150,000, a very competitive price, but no customer interest was forthcoming. The Saab 105D was a subsequent, preliminary project study which featured extended wingtips and other aerodynamic improvements.

When the Saab 105C was launched in 1966, the aircraft was firmly established as a military-spec trainer, and it could no longer be simply remodelled as an executive jet. By that time the first Gates Learjets and Dassault Falcons had flown, and the 105C was not a serious competitor.

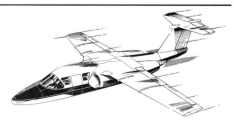

Saab Sk 60A

The first production Sk 60A flew on 27 August 1965, and all 150 deliveries to the Flygvapnet, bar one (the prototype Sk 60C), were of Sk 60As. Deliveries totalled 30 during 1966, 57 in 1967, 59 in 1968 and the last four in 1969. In trainer configuration two pilots are carried on side-by-side Saab ejection seats. The rear-hinging canopy is made from double-curved acrylic/glass. The Saab 105 was initially powered by two 6.8-kN (1,540-lb) Turboméca Aubisque (RM9) ducted fan engines with a bypass ration of 2:1. The engines were later uprated to 7.3 kN (1,640 lb) as the RM9B. The aircraft is capable of operating from a 600-m (1,970-ft) runway and can climb to 10000 m (32,800 ft) in less than six minutes. The shoulder-mounted wing provides 1.4 m (4½ ft) of ground clearance, and combines one-piece stressed skin construction with two continuous spars. It is fitted with single-slotted flaps and air brakes. Fuel is housed in two integral fuselage tanks and two wing tanks with a total capacity of 1400 litres (310 Imp gal). Armament could be accommodated on six underwing hardpoints. However, although all Sk 60As can be armed, in practice the attack aircraft need further specialist equipment. Today, the number of Sk 60As remains fixed at 56 (seven having been lost in accidents). It is far easier to convert aircraft to four-seat configuration, and this is done on a regular basis, although they remain in 'temporary Sk 60E' configuration for only 200 hours at a time.

Right: An Sk 60A (nearest the camera) formates with an unarmed Sk 60C. Sk 60As operate permanently without underwing pylons and all serve with the basic flying training school at F5.

Left: An early production Sk 60A (600003), wearing the markings of the Försökcentralen (central research establishment), turns in on finals. This unit conducted extensive trials with the first four Sk 60s.

Above: The first production Sk 60 takes off for its maiden flight from Saab's Linköping home on 27 August 1967. Nearly 30 years later this aircraft is still in regular service, training students at Ljungbyhed.

Saab Sk 60B

The 1966-67 Flygvapnet budget included special funds to modify existing Sk 60s to carry the Saab 305A (Rb 05) ASM. These aircraft were allocated the initial, provisional designation of A 60 and were the world's first light jet trainers to be thus equipped. Some Sk 60As were subsequently modified for the light attack/reconnaissance mission as Sk 60Cs. However, from the outset all Sk 60s could be equipped with underwing pylons and weapons, and aircraft with these pylons are designated Sk 60B. They have also been fitted with the Ferranti F-105 ISIS (Integrated Strike and Interception System) gunsight, but otherwise are virtually identical to the unarmed Sk 60A trainers. Approximately 45 Sk 60s are permanently fitted with these pylons, gun camera brackets, armament panel and electrical systems. Another small clue to their identity can be found on the nose, where a dielectric panel (associated with the guidance system for the Rb 05, though this is not actually fitted) causes the wing number to appear slightly higher than on Sk 60As.

Saab Sk 60B

Blade antenna now fitted to some Sk 60s

Aircraft is identical to Sk 60A but for three hardpoints with larger central pylon under each wing

Ferranti F-105 ISIS weapons sight on pilot's side (port) only

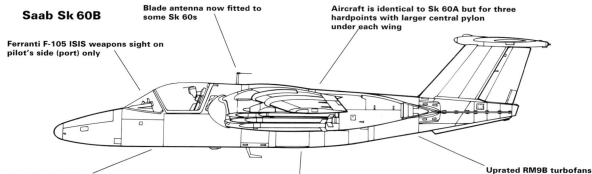

Provision for Rb 05 missile guidance system, though never fitted to service aircraft

Uprated RM9B turbofans

13.5-cm rockets (Arak) usually carried in pairs

The first dedicated Sk 60B Division, with 15-20 aircraft, was formed at Luleå, as part of F21 in early 1973. The light attack squadron (3 Division, Urban gul) operated alongside the unit's two Draken squadrons – one with J 35F interceptors and one with S 35E recce aircraft. This made F21 the first Flygvapnet unit to simultaneously operate fighter, ground-attack and reconnaissance components. The Sk 60 squadron disbanded in 1982, becoming a JA 37 unit, its Sk 60s transferring to F20 (later F16) at Uppsala. In wartime the Flygvapnet would field four light attack squadrons of Sk 60B/Cs, dispersed in northern Sweden and crewed by instructors from F5 and F16 (approximately two squadrons from each wing), and by the Lansen pilots from F16M.

Left: The Sk 60 complies with CAR Pt 3 Aerobatic and Utility standards. A weapon load of 700 kg (1,543 lb) can be carried in excess of the max aerobatic weight of 3800 kg (8,380 lb).

Above: The Sk 60B can carry up to 12 13.5-cm rockets and has also been cleared to carry a range of MATRA rocket pods and bombs in addition to Swedish weapons.

Saab Sk 60C

As the Viggen, 'System 37', evolved, so too did the air force's operational doctrine. It soon became obvious that there would never be enough Viggens to go around. The Viggen was a sophisticated and expensive aircraft, and it would be necessary to find a lightweight and affordable attack aircraft for some secondary close air support, FAC and reconnaissance duties.

The Saab 105 was an obvious choice, already being cleared with hardpoints and a range of air-to-ground weapons. Chief among these are the FFV (MATRA) SA-10 30-mm ADEN cannon pod, Bofors 13.5-cm high-explosive and 14.5-cm armour-piercing rockets and Saab Rb 05 ASMs. Radius of action on a lo-lo-lo mission with a warload of 1000 kg (2,204 lb) is 400 km (248 miles).

60010, the first Sk 60C, was the only Sk 60 not built as an Sk 60A – all the other

Sk 60Cs were conversions. It made its first flight on 18 January 1967 and was temporarily fitted with a box-like pod on the port wingtip containing a test camera to record rocket firing and bomb release. Twenty-seven Sk 60Cs (two have been written off) serve with F5 and F16.

The Sk 60C's most obvious feature is its extended nose housing a Fairchild KB-18 panoramic camera (Swedish designation SKA 29) with a 76-mm, f/2.8 lens. The KB-18 is depressed to 20°, and has a field of view of 180° x 40°. It uses 70-mm film and is capable of night photography when

combined with a Chicago Aerial flash pod. The Sk 60C is intended to operate in a low-level environment, and the camera is effective down to heights of 50 m (164 ft). The Sk 60C's reconnaissance role is limited today, and the aircraft are chiefly employed as light attack aircraft.

Saab Sk 60C side view

Extended nose housing Fairchild KB-18 panoramic camera

14.5-cm rockets are only carried singly, usually on innermost and outermost stations

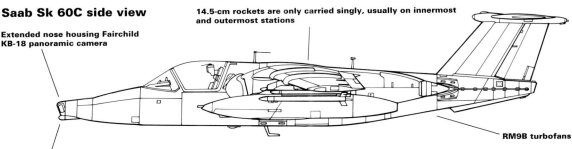

RM9B turbofans

Disused missile (Rb 05) guidance antenna

Some Sk 60Cs have camera removed and windows faired over.

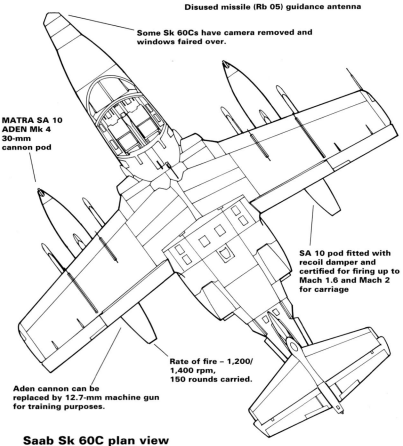

MATRA SA 10 ADEN Mk 4 30-mm cannon pod

SA 10 pod fitted with recoil damper and certified for firing up to Mach 1.6 and Mach 2 for carriage

Rate of fire – 1,200/1,400 rpm, 150 rounds carried.

Aden cannon can be replaced by 12.7-mm machine gun for training purposes.

Saab Sk 60C plan view

Above: The prototype Sk 60C was temporarily fitted with a wingtip camera pod to record weapons tests. The pristine SA 10 cannon pods seen here look completely unused thus far.

Below: The Sk 60C's 'old world' cockpit is representative of all Swedish Sk 60s with some small differences. Chief among these is the ISIS sight above the coaming.

Saab Sk 60D

Flygvapnet specification for the Saab 105 set many demands, one of which was that the aircraft should be easily configurable for VIP/liaison/transport duties. By removing the ejection seats the cockpit can be configured to carry four fixed airline-style seats. In fact, two types of seat are available – an upholstered version, and a more austere seat which allows the occupant to wear a parachute and the associated rescue/survival gear. The cockpit was equipped with ADF1/2 navigation systems, FR17 and Collins VHF 20 radios. Instruments were calibrated in 'metric'. Ten Sk 60Ds were permanently converted for Swedish service

between 1976/1977, and three remained operational in June 1995. A standard Sk 60A can be reconfigured for transport duties in about 90 minutes. With four passengers it has a range of 1580 km (982 miles). Flygvapnet Sk 60Ds do not carry external fuel tanks. Along with the Sk 60Es, the Sk 60Ds will be retired within the next few years and used for spares support.

This Sk 60D, wearing the marks of now disbanded F18, was one of a very small number of Sk 60s to wear this splinter-style camouflage, as opposed to the more subtle standard brown/green finish, with grey undersides.

Saab Sk 60E

Between 1978 and 1984, Sk 60E navigation trainers were operated by F5, with some instructors from SAS, to train commercial pilots, who would also become reserve officers, rather than have the airline poach permanent air force pilots (who are free to leave at any time as they have no formal contracts). The Sk 60E was fitted out with commercial instruments, such as VOR, ADF, FR17 and FR 33 radios and Sperry C-14A twin-platform gyrocompass. Instruments were calibrated in 'feet' and 'knots'. Three aircraft were thus modified between 1976/77. They were then the only Sk 60s to be fitted with ILS. The cabin was configured for an instructor and three students. Today, the Sk 60Es are operated as Sk 60D

transports. F5 maintains a commercial school in association with Swedish airlines, using the Sk 61 Bulldog and Tp 54 (PA-31-350), and staffed by military and civilian personnel. The Sk 60Ds (and Es) will not receive the FJ44/RM15 upgrade and, despite their low hours, will be retired by the year 2000.

The Sk 60E was intended as an airways-equipped civilian trainer and served in that role during the rapid airline recruiting days of the late 1970s and early 1980s. Today, the Sk 60Es and Ds are virtually interchangeable. They all are slated for retirement in the near future. The lack of ejection seats is obvious in this view.

Saab Sk 60 FJ44 upgrade – Sk 60(W)

On 4 November 1993 the Flygvapnet signed a contract with Williams/Rolls to cover the re-engining of a total of 115 Sk 60A/B/Cs with FJ44-1C turbofans (Swedish designation RM15). A total of 137 Flygvapnet Sk 60s (138 according to Saab) are currently active, with the 13 Sk 60D/Es due for imminent retirement. All the Sk 60Bs and Cs (27 and 45 respectively) will be re-engined, along with 43 Sk 60As. Options are held on re-engining the small number of Sk 60As that will remain after the initial batch of 115 aircraft have been completed. Either way, the upgrade programme is scheduled to be completed by mid-1998, and the upgraded Sk 60s will remain in service until 2010-2015.

The new engine was one of the few available that would fit in the Sk 60. Integration studies were also undertaken on the Turboméca Larzac, but the FJ44 offered the minimum of airframe changes. The FJ44 installation will reduce the aircraft's take-off weight by 500 kg (1102 lb). The engine's

bypass ratio is 3.28:1 compared to the RM9B's 2.15:1, and it weighs 210 kg (463 lb) compared to 300 kg (661 lb). The Sk 60's noise footprint will be reduced from 105 dBa to 85 dBa, at a distance of 15 m (33 ft). The only external difference between 'old' and 'new' Sk 60s will be slightly larger engine covers/access doors on the latter. A new digital engine and fuel control system has been fitted. Flygvapnet sources state that the converted Sk 60s will not be allocated a new designation, but Saab are already referring to the aircraft as the Sk 60(W).

The first Sk 60 to be modified was 60072, which made its maiden flight in August 1995. The first 10 conversions (including the prototype) are being wholly undertaken at Saab's Linköping factory. Between 1996 and 1998, 105 subsequent aircraft will be disassembled at Ljungbyhed and their fuselages will be trucked to Linköping, where they will be fitted with the new fuel/hydraulic system and will have the

rear inlet section modified. They will then be transferred back to F5, where the engines will be fitted, the wings replaced and test flying undertaken. 60072 has been transferred to the FMV test unit (FMV:Prov) at Malmen, for a test flying programme involving four pilots from FMV, Saab and F5. This will continue until the end of the year, and is intended to validate extensive computer modelling of the re-engined Sk 60 undertaken by Saab, with assistance from BAe. The second aircraft will be delivered to F5 in April 1996.

The decision on which aircraft to upgrade was made according to their position in the regular maintenance cycle. F5 performs four levels of check on its Sk 60s: E (every 400 hours), F (800 hours), G (1,600 hours) and H (the 3200 hour major overhaul). Aircraft undergo checks in the order E, F, E, G, E, F, E, H. Hours are balanced across the fleet by swapping wings from one aircraft to another, from an Sk 60E to an A, for example.

The first FJ44-powered Sk 60 (an Sk 60B) takes shape at Linköping during early 1995.

An avionics upgrade has not been incorporated in the upgrade contract. The air force hopes that one can be initiated in the near future and hopes to incorporate two cockpit MFDs, but perhaps not a HUD. Replacement of the troublesome 1963-vintage flight instruments, which have been blamed for some crashes, is most important.

Saab 105XT

Saab completed development of the first major export variant as first deliveries of the Sk 60 to the Flygvapnet began. Intended as multi-role trainer/light attack/liaison aircraft with secondary target-towing and reconnaissance capabilities, the 105XT was designed primarily for export to hot-and-high nations. The XT in the designation stands for eXport Tropics. The type was essentially a re-engined Sk 60B, powered by a pair of non-afterburning 2,850-lb (12.68-kN) General Electric J85-17B (CJ-610) turbojets. These engines had nearly 75 per cent more power than the Sk 60's tiny Aubisque powerplants. This increased range and payload accordingly. Saab promoted the 105XT as being capable of carrying up to five (rather cramped) passengers, or a stretcher and attendant with all but the pilot's seat removed. Equivalent Swedish Sk 60Ds have never flown in any of these configurations. Compared to the Sk 60A's MTOW of 4500 kg (9,920 lb), the 105XT weighed in at 6416 kg (14,146 lb). To offset the new powerplant's higher fuel consumption, internal capacity was increased to 2050 litres (451 Imp gal). The XT prototype (the modified and re-engined second prototype SK 60, SE-XBZ) first flew on 29 April 1967, for 38 minutes,

The Saab 105XT demonstrator went through a series of incarnations. This is the aircraft seen fairly early in life with 250- and 500-kg bombs.

in the hands of deputy chief test pilot K. E. Fernberg. Much climatic testing of the new powerplants was undertaken, including cold soak/cold start testing during the winter of 1967/68. In 1968 a new servo-operated aileron was fitted to alleviate the higher stick forces encountered at the XT's higher airspeed. Vibration and flutter testing followed during summer 1968. Special impulse rockets on the wings were fired when the aircraft was travelling at 900-1000 km/h (560-621 mph) in a dive. The resultant vibrations were recorded on an oscilloscope in the cockpit. A nose probe was added in 1969, and in this configuration SE-XBZ was evaluated by the Swiss air force. Once the integrity of the wing design had been confirmed, in August 1973, the aircraft underwent weapons testing at Vidsel in northern Sweden, wearing the temporary military code '43'.

In October 1968 the XT was flown with combinations of up to 12 rockets or gun pods, for rocket plume and gun gas ingestion tests. Winter came early that year and weapons testing proceeded alongside

further cold weather trials. For example, the fully armed Saab 105XT was left overnight in temperatures of -10°C. It started first time on the internal battery and the guns were successfully fired immediately afterwards.

At the 1970 Hannover air show the Saab

105XT demonstrated its new-found target-towing capability, carrying the US-designed Del Mar target system. This comprised a DX-5A tow reel, DL-45 launcher and DF-4MFC airborne gunnery target. The Saab 105XT was flown with the full complement

of Swedish air-to-ground ordnance and also with early model Sidewinder (or Rb 24 in Swedish parlance) AAMs. Other options included the Bullpup and AS20 ASMs.

Four of the six pylons (the outermost and innermost) were stressed to accept 610 kg (275 lb), while the remaining (centre) pair could carry up to 1000 kg (450 lb). With a maximum load of six 500-lb bombs the Saab 105XT had a radius of action of 350 km (217 miles), on a lo-lo-lo mission with 2.5 minutes over the target and 10 per cent reserves. On a hi-hi-hi mission, this increased to 850 km (528 miles).

The Saab 105XT also undertook much of the flight test clearance programme for the reconnaissance pod designed for Austria's

Saab 105Ös. Saab drew up plans for a dedicated 105XT recce version. The basic airframe resembled the Sk 60C, with the latter's panoramic camera nose. The new aircraft would also have carried a sideways-looking airborne radar (SLAR) in two fairings running along the side of the forward fuselage, with the additional new electronic equipment required fitted behind the two pilot's seats, and a Doppler radar altimeter under the tail. The aircraft attracted no export interest, and the proposal was never proceeded with.

Over the course of its life the Saab 105XT prototype appeared in various guises. It first flew in the natural metal finish inherited from its days as the second Saab

In its definitive form the Saab 105XT was a sizeable improvement over existing aircraft, boasting much improved handling and warload.

105 prototype, albeit with red and blue stripes, 'Saab 105XT' titles and a large Swedish registration on the fin and under the wings. As weapons tests commenced, SE-XBZ adopted a two-tone green camouflage on its top surfaces and fuselage sides with a grey finish on the undersides. Later the civilian registration was temporarily replaced by the military code '43', with natural metal fin and rear fuselage, a new brown/green camouflage with grey undersides and a nose-mounted air-data probe.

Between 1974 and 1976 SE-XBZ became a testbed for a ground-moveable wing leading edge intended to find the best wing profile for the Saab 105XH proposal for Switzerland. At a setting of 2° this gave an improved 'peak transonic profile', compared with the standard NACA 64 wing profile. Dubbed the Saab 105-2XT, SE-XBZ retained the instrumentation probe on the nose and

flew with multiple weapons loads to validate the new wing design, with its drooped leading edge.

The 105-2XT's leading-edge extension was ground-adjustable through settings of 0°, 2°, 15° and 20°. By the beginning of February 1971 the aircraft had made 30 test flights and after 60 to 65 flights the angle was fixed at 2°, as the best compromise. This modification also entailed extending the leading edge by 7 per cent of the mean chord and installing a new wing section. The resultant change in pressure distribution increased maximum wing loading at high speeds. In practical terms this more than doubled the amount of *g* the Saab 105XT could sustain at Mach 0.80, from 2.5 to 7 *g* (clean). The operational maximum was 6 *g*, but this still allowed for a fatigue life of 1,500 hours for an attack aircraft, or 3,000 for a trainer. This has been the only alternative wing design actually flown on the Saab 105.

In July 1970 Saab began deliveries of the basic Saab 105XT to the Austrian air force under the designation Saab 105Ö. The Saab 105XT prototype went on to serve as the Saab 105G trials aircraft.

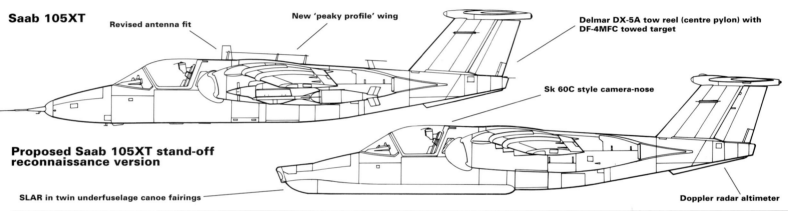

Saab 105XT
Revised antenna fit
New 'peaky profile' wing
Delmar DX-5A tow reel (centre pylon) with DF-4MFC towed target
Sk 60C style camera-nose

Proposed Saab 105XT stand-off reconnaissance version
SLAR in twin underfuselage canoe fairings
Doppler radar altimeter

Saab 105Ö

In the summer of 1968 the Austrian Ministry of Defence ordered a batch of 20 Saab 105XTs, designated 105Ö (for Österreich – Austria), to replace the survivors of 30 J 29F Tunnen ('flying barrels') delivered in 1960/61. An option for a further 20 was also taken out and fulfilled in July 1969. Saab 105Ös were delivered to Austria's Heeresfliegerkräfte (Army Aviation Force) and the 40 aircraft cost Sch 800 million (£12.9 million).

Saab 105Ö deliveries to the air arm – now known as the Österreichische Luftstreikräfte (ÖLsk), Austrian air force – began in July 1970. Nine aircraft arrived in 1970, 18 in 1971, and the last 13 in 1972. This allowed the retirement of the Vampire (on 20 April 1972) and the Saab J29 (on 21 June 1972). The new Saabs were operated by the Überwachungsgeschwader (surveillance wing), of Fliegerregimenter II, based at Graz-Thalerhof, and by the Jagdbombergeschwader (fighter-bomber wing), of Fliegerregimenter III, at Linz-Hörsching. Each wing eventually possessed two staffeln (squadrons) with seven or eight aircraft. The Graz-based aircraft were charged with air defence, though they were left standing by the occasional Learjet that it was their duty to shepherd out of Austrian air space. One of

Saab 105Ö
Improved avionics and ECS fit
Additional fuel tanks in wing
Vicon camera pod, carried to port
General Electric J85 turbofans

the Linz-based squadrons was the air force weapons training/close support unit and regularly practiced at the Allensteig gunnery range, north of Vienna, and at the Seetaler bombing range in the Alps. Between 1975 and 1984 a six-ship, Saab 105-equipped aerobatic team – first the 'Silver Birds' then, from 1976, 'Karo As' – was flown by air force instructors.

Austrian Saab 105s carry 500 kg (1,102 lb) of additional fuel in a reinforced wing with servo ailerons and are fitted with a new cooling turbine to improve the cockpit air system at the higher operating altitudes and speeds encountered by the J85-powered

aircraft. All these improvements were first trialled by Saab on their Saab 105XT demonstrator. The Austrian aircraft also possess a more extensive avionics fit than the Sk 60, but the most substantial difference between them and Swedish-flown aircraft is their podded reconnaissance system. The J 29Fs could carry a Vinten photo-reconnaissance pack and it was important that this capability should be transferred to the Saab 105s.

As carried by the Saab 105Ö, the recce pod houses (a maximum of) one forward-facing and four downward/oblique-facing Vinten 70 cameras. These range from 304-mm f/4 cameras for daytime photography to 76-mm, f/2.8 lenses for low-level night-time missions. Systems installation in the parent aircraft (chiefly that of the hot air supply for the pod) forced the pod to be carried on the port outboard pylon, while a flash unit could be carried on the starboard, inner pylon.

The Austrian 105s carried a limited selection of weapons, and, despite the close support tasking, never carried bombs. The

most common load for air policing is four rockets, although a theoretical maximum of 12 can be carried in the close support role. Blue-painted Bofors 6.3-cm rockets were used for training, with a motor but no warhead. The 7.5-cm Frida rocket was not only for training, but also for air-to-air and air-to-ground uses. Operational rockets were painted green, and inert, motorless ground-crew training rounds were black. The 6.3-cm rocket was intended as a training round for the 13.5-cm Gerda rocket, but this was not procured. The BL755 was tested during 1980, but was similarly not procured.

The Saab 105 was selected as an 'interim' combat type, as the J 29s so urgently needed replacement, while also serving as a replacement for Vampire T.Mk 55 and Fouga Magister trainers. However, while Austria's lengthy search for a new front-line type became almost legendary, the Saab 105s were the ÖLsk's sole combat aircraft until the arrival of the first of 24 former Swedish J 35Ö Drakens in 1988. The Drakens are expected to reach the end of their useful lives in 1995/96, and the Saab 105Ös may well outlast them in Austrian service. Currently, 29 Saab 105Ös remain in service. After a major air force reorganisation on 1 July 1995, all aircraft report directly to a single regiment, based at Linz-Hörsching, which loans about eight aircraft to the Überwachungsgeschwader at Zeltweg and Graz at any given time.

This Saab 105Ö wears the cartoon tiger badge of the former 1 Staffel (yellow), based at Hörsching and is carrying a load of 7.5-cm Frida rockets. This aircraft was painted with tiger stripes in 1993.

Saab 105XH

In June 1970 Saab forwarded a proposal to the Swiss government to supply the air force with 120-150 Saab 105s, as dedicated ground attack aircraft, designated Saab 105XHs. The XH was based on the Saab 105XT, but featured several radical new features, including more internal fuel, advanced avionics, better handling, and an increased take-off weight. The most substantial differences were the inclusion of an internal gun and the addition of wingtip fuel tanks.

Switzerland's 10-year search for a close-air support aircraft was finally resolved by the acquisition of refurbished Hunter FGA.Mk 58s and by the 1976 'Peace Alps' purchase of 66 Northrop F-5Es. However, in 1970 the (then) LTV-Aerospace A-7 Corsair was selected after a Swiss technical evaluation. This decision was overturned by the Swiss parliament, who which an alternative to the American aircraft. Saab stepped in with the 105XH, which marked a significant departure from the type's simple trainer origins.

The primary armament of the 105XH was intended to be a 30-mm ADEN cannon, housed in a fairing under the port fuselage. This gun was to be fed by a 100-round, pre-loaded magazine, housed internally, beside the gun, to starboard. Fitting an internal weapon freed up space on the pylons, but also cut back on the aircraft's fuel load. As a result, fixed 200-litre (44-Imp gal) tip-tanks were added. The six underwing pylons

were restressed to carry slightly higher loads, and the Saab BT9R precision bombing system and laser-rangefinder was to be carried in the nose. Further improved avionics came in the form of a Doppler radar, Ferranti ISIS gun sight, a roller map display and optional HUD (the latter derived from the Viggen's HUD). The reprofiled wing was to be fitted with a drooped leading edge and maximum take-off weight was pegged at 7000 kg (15,430 lb), an increase of 500 kg (1,102 lb) over the armed Sk 60B. As a final touch, a brake parachute was also included in the unsuccessful design.

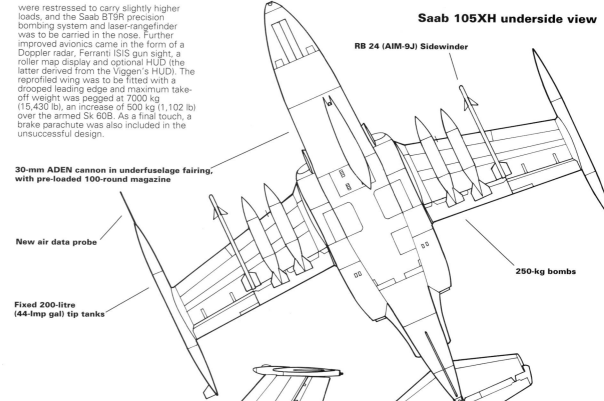

Saab 105XH underside view

RB 24 (AIM-9J) Sidewinder

30-mm ADEN cannon in underfuselage fairing, with pre-loaded 100-round magazine

New air data probe

Fixed 200-litre (44-Imp gal) tip tanks

250-kg bombs

Saab 105XH side profile

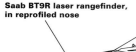

Saab BT9R laser rangefinder, in reprofiled nose

Decca Doppler radar altimeter

Saab 105G

The Saab 105G built on the experience gained with the Saab 105XT, incorporating much of the improved manoeuvrability planned for the 105XH, with increased weapons load, provision for drop tanks and enhanced avionics. It retained the higher powered J85 engines of the XT, in J85-GE-17B form, but its most important feature was a new avionics suite, adapted from that of the Lansen, with a precision nav/attack system. This included a Decca Doppler radar altimeter and roller map display, an Elliott air data computer, Sperry twin gyro INS, and the Ferranti ISIS sight. A Saab BT9R marked target seeker and laser rangefinder (with an associated ballistic computer) was fitted in the reprofiled nose. An obvious outward modification to the Saab 105G would have been a modified wing leading edge which would have allowed it to attain almost double the g-load at high speed without increasing drag or degrading low-speed handling. The wing chord was to have been extended by 5 per cent. The air brakes would have been enlarged by 60 per cent and the maximum flap deflection was to have increased to improve turn performance and short field landing capability.

The integrated weapons delivery system was optimised for use at distances of up to 6 km (3.7 miles), from a dive angle of 60° at speeds of up 590 kt (1050 km/h; 652 mph). The Saab 105G was stressed to carry up to 2350 kg (5,180 lb) of ordnance, but was also

touted for training, reconnaissance, target tug and transport duties.

SE-XBZ served as the Saab 105G demonstrator, having previously acted as the Saab 105XT prototype. It retained the fixed, drooped leading edge developed for the Saab 105XH, and lacked the other aerodynamic changes intended for the 105G. The type débuted at the 1972 Farnborough show and appeared at the following year's Paris show. The aircraft, together with a Saab 105O borrowed from the line (and temporarily coded '86') conducted a series of weapons trials during the spring of 1972, but failed to find customers in a market that was becoming saturated with similar types. The 105G remained as a trials aircraft with Saab until it retired to the air force museum in 1992, where it still resides, in the eye-catching yellow and blue scheme adopted in 1976.

The 105G was the final development of Saab's diminutive design – an advanced and capable light attack aircraft that was never seriously allowed to compete in the international market. Here, the prototype is carrying a Red Baron recce pod.

Saab 105G

Ferranti ISIS sight and roller map display in cockpit

Airbrakes enlarged by 60 per cent

J85-17B turbofans

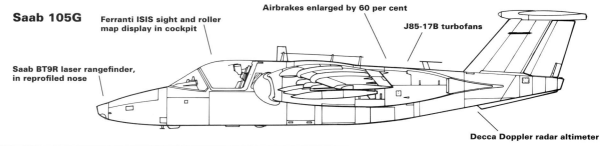

Saab BT9R laser rangefinder, in reprofiled nose

Decca Doppler radar altimeter

Saab 105S

The Saab 105S was a version of the 105XT, tailored for Finland's mid-1970s requirement for a jet trainer to replace its Fouga Magisters by 1987. Saab had already gained a foothold in Finland, having supplied the

Ilmavoimat with the S 35XS Draken. By July 1976 the field had narrowed to the Saab 105S and British Aerospace Hawk, though at one time such types as the revamped Aérospatiale Fouga 90, JuRom (IAR 93) Orao and (improbably) the Vought TA-7C had been under consideration.

Both UK and Swedish manufacturers

were offering generous industrial offsets, to the value of 100 per cent of the 50-aircraft requirement. Saab proposed that Valmet would build 50 per cent of the airframe and assemble the General Electric J85 engines. The Saab 105S was cheaper than the Hawk, with a price quoted at £1.07 million as opposed to £1.26 million. A Saab 105O was

borrowed from the Austrian air force (prior to delivery) and demonstrated, in Finland, in the first half of 1972. In the event, the BAe Hawk was selected and another of the very few countries on Saab's restricted list of potential export customers slipped away.

All front-line Flygvapnet units regularly practise dispersed operations. The Sk 60 is no stranger to this environment either, as it undertakes advanced weapons training and fully-fledged attack missions with F5 and F16.

Sk 60 in Swedish service

On 6 March 1964 the Swedish government formally authorised the production of 130 new jet trainers for the Flygvapnet, as the Sk 60. Saab had already started production on the strength of a letter of intent. A further 20, equipped as dedicated light attack aircraft, were ordered in 1965. Deliveries to the Försökcentralen test unit began in December 1965, and were followed by delivery of the first operational Sk 60As to F5 (the flying training wing) at Ljungbyhed, in the spring of 1966. Production continued at the rate of six per month.

Swedish pilots were initially screened on the Saab Sk 50 Safir (and later the Sk 61 Bulldog) for 30 hours before transitioning to the Sk 60 for a further 105-130 hours. This compared with the previous system of 75 Sk 50 hours followed by 85 hours on the elderly Sk 28 Vampire. Having left the Sk 60, new pilots soloed in the two-seat Sk 35C Draken after only 35 hours. By March 1970 Sweden's Saab 105s had accumulated 50,000 flying hours in 60,000 sorties. By 1976 the total was over 170,000 hours.

By 1977 the Sk 60 was at the height of its career, with the Flygvapnet transitioning from Draken to Viggen. Students who completed their Sk 60 course at F5's Krigsflygskolan (Royal War Flying School) passed on to the Royal Air Force Academy/War School (F20), at Uppsala, which used Sk 60B/Cs for weapons training. Luleå-based F21 operated a squadron of attack-configured Sk 60Bs and Sk 60Cs, while target-towing Sk 60A/Bs were in use with F13M at Malmen. A handful of four-seat Sk 60s served with the Staff Liaison Flight of F18, at Stockholm/Tullinge. Virtually all Sk 60s had a secondary attack role in time of war, but only F20 and F21 trained for close air support missions. F21 was intended to provide the nucleus for the Lätt Attackenheter (light-attack units) that would be formed in the event of war. In 1982 F20 was disbanded, transferring its aircraft to F16 at the same base. F21 swapped its Sk 60s for Viggens, and its aircraft also went to F16.

Today every Sk 60 wears the codes of F5, and is allocated from a central maintenance pool. All basic flying training is undertaken by F5, while F16, at Uppsala, handles advanced and weapons training. Every air force wing (Flygflottilj) has a small number of Sk 60s attached for continuation training and liaison duties. F5 is formally known as the Krigsflygskolan (military flying school) and its main operational arm is the GFU (Grundläggande Flygutbildning/basic flying training school).

Until recently, F5 was unique in the Swedish air force as it had four flying squadrons, numbered one to four. Usually, a wing with four flying squadrons (Divisioner) is numbered one (röd/red), two (blå/blue), three (gul/yellow) and five (svart/black). The

Sk 60 – Svenska Flygvapnet

Serial	Variant	Unit code	Notes
60001	Sk 60C	FC-36, F5-01, F21-10, F5-01	
60002	Sk 60C	FC-37, F5-02, F21-11, F5-02	
60003	Sk 60C	FC-38, F5-03, F20-03, F21-12, F5-03	
60004	Sk 60C	FC-30, F5-04, F21-17, F5-04	
60005	Sk 60C	F5-05, F20-05, F5-05	
60006	Sk 60C	F5-06, F20-06	w/o 27/2/1981
60007	Sk 60C	F5-07, F21-13, F5-07	
60008	Sk 60A	F5-08	w/o 25/5/1967
60009	Sk 60C	F5-09, F21-21, F5-09	
60010	Sk 60C	-42, F5-42, FC-32, F5-10, F21-20	
60011	Sk 60C	F20-11, F21-06, F5-11	
60012	Sk 60C	F5-12, F21-14	
60013	Sk 60C	F5-13, F20-13, F21-15, F5-13	
60014	Sk 60C	F5-14	
60015	Sk 60C	F5-15, F20-15, F5-15	
60016	Sk 60C	F5-16	
60017	Sk 60C	F5-17, F21-16, F5-17	
60018	Sk 60C	F5-18	
60019	Sk 60C	F5-19	w/o 4/6/1975
60020	Sk 60C	F5-20	
60021	Sk 60C	F5-21, F21-17, F5-21	
60022	Sk 60C	F5-22, F20-22, F5-22	
60023	Sk 60C	F5-23, F20-23, F5-23	
60024	Sk 60C	F5-24	
60025	Sk 60C	F5-25, F20-25, F21-15, F5-25	
60026	Sk 60C	F5-26, F20-26, F21-19, F5-26	
60027	Sk 60C	F5-27	
60028	Sk 60C	FC-38, F5-28	
60029	Sk 60A	F5-29	w/o 6/5/1968
60030	Sk 60A	F5-30	w/o 7/5/1992
60031	Sk 60C	F5-31	
60032	Sk 60C	F5-32	
60033	Sk 60A	F5-33, F21-25, F5-33	Team 60
60034	Sk 60B	F5-34, F20-34, F5-34	
60035	Sk 60B	F5-35, F20-35, F5-35	
60036	Sk 60B	F5-36, F21-09, F21-07, F5-36	
60037	Sk 60B	F5-37, F20-37, F5-37	
60038	Sk 60B	F5-38, F21-08, F5-38	
60039	Sk 60B	F5-39, F21-09, F5-39	
60040	Sk 60A	F5-40	
60041	Sk 60B	F5-41, F20-41, F5-41	
60042	Sk 60A	F5-42, F20-42, F5-42	
60043	Sk 60A	F5-43, F21-26, F5-43	w/o 16/3/1989
60044	Sk 60A	F5-44, F21-01, F5-44	
60045	Sk 60A	F5-45	
60046	Sk 60A	F5-46	
60047	Sk 60A	F5-47	
60048	Sk 60A	F5-48	
60049	Sk 60B	F5-49	
60050	Sk 60B	F5-50	
60051	Sk 60A	F5-51, F5-31, F5-51	
60052	Sk 60B	F5-52, F20-52, F5-52	
60053	Sk 60A	F5-53	
60054	Sk 60B	F5-54, F20-54, F5-54	
60055	Sk 60B	F5-55, F20-055, F20-55, F5-55, F20-55	
60056	Sk 60B	F5-56	
60057	Sk 60B	F5-57, F21-02, F5-57	w/o 11/10/1988
60058	Sk 60B	F5-58, F20-058, F20-58, F5-58	
60059	Sk 60B	F5-59	
60060	Sk 60B	F5-60	
60061	Sk 60A	F5-61	Team 60
60062	Sk 60A	F5-62	Team 60
60063	Sk 60B	F5-63, F20-63, F5-63	
60064	Sk 60B	F5-64, F21-03, F5-64	
60065	Sk 60B	F5-65	
60066	Sk 60B	F5-66	
60067	Sk 60B	F5-67	
60068	Sk 60B	F5-68, F21-04, F5-68	
60069	Sk 60B	F5-69, F20-69, F5-69	
60070	Sk 60B	F5-70	
60071	Sk 60B	F5-71	
60072	Sk 60B	F5-72, F21-05, F5-72	
60073	Sk 60B	F5-73	
60074	Sk 60B	F5-74, F20-74, F5-74	
60075	Sk 60B	F5-75	
60076	Sk 60B	F5-76, F90-76, F5-76	
60077	Sk 60B	F5-77, F21-06, F5-77	w/o 19/5/1994
60078	Sk 60A	F5-78, Team 60	w/o 11/9/1980
60079	Sk 60A	F5-79	
60080	Sk 60B	F5-80	
60081	Sk 60A	F5-81	
60082	Sk 60B	F5-82, F20-82, F10-82, F5-82	
60083	Sk 60B	F5-83, F21-07, F5-83	
60084	Sk 60A	F5-84	
60085	Sk 60B	F5-85, F20-85, F5-85	
60086	Sk 60B	F5-86, F20-86, F21-08, F5-86	
60087	Sk 60B	F5-87, F20-87, F5-87	
60088	Sk 60A	F5-88	
60089	Sk 60B	F5-89, F20-89, F5-89	
60090	Sk 60B	F5-90	
60091	Sk 60B	F5-91	
60092	Sk 60A	F5-92	
60093	Sk 60A	F5-93, F20-93, F5-93	
60094	Sk 60A	F5-94, F20-94, F5-94	
60095	Sk 60A	F5-95	
60096	Sk 60A	F5-96	Team 60
60097	Sk 60D	F5-97, FC-30, F5-97, FC-37	
60098	Sk 60A	F5-98	Team 60
60099	Sk 60A	F5-99	
60100	Sk 60A	F5-100, F20-100, F5-100	
60101	Sk 60A	F5-101	
60102	Sk 60A	F5-102	
60103	Sk 60A	F20-103, F5-103, FC-33, F5-103	
60104	Sk 60A	F5-104	
60105	Sk 60A	F20-105, F5-105	
60106	Sk 60A	F20-106, F5-106	
60107	Sk 60A	F5-107	
60108	Sk 60A	F20-108, F5-108	
60109	Sk 60A	F20-109, F20-109, F5-109	
60110	Sk 60A	F20-110, F5-110	
60111	Sk 60A	F20-111, F5-111	
60112	Sk 60A	F20-112, F5-112	
60113	Sk 60A	F20-113, F5-113	
60114	Sk 60A	F5-114	
60115	Sk 60A	F20-115, F5-115	
60116	Sk 60A	F5-116	
60117	Sk 60A	F20-117, F5-117	
60118	Sk 60A	F5-118	
60119	Sk 60A	F5-119, F20-119, F5-119	
60120	Sk 60A	F5-120	
60121	Sk 60A	F5-121	w/o 31/10/1974
60129	Sk 60A	F5-127	
60123	Sk 60A	F5-123, F20-123, F5-123	Team 60
60124	Sk 60A	F5-124	
60125	Sk 60A	F5-125	Team 60
60126	Sk 60A	F5-126	
60127	Sk 60A	F5-127	
60128	Sk 60A	F5-128	
60129	Sk 60A	F20-129, F5-129	
60130	Sk 60A	F20-130, F5-130, F21-27, F5-130	
60131	Sk 60D	F5-131, F18-131, F5-131	
60132	Sk 60D	F5-132	
60133	Sk 60A	F5-133	
60134	Sk 60A	F20-134, F5-134	Team 60
60135	Sk 60A	F20-135, F5-135	
60136	Sk 60A	F20-136, F5-136	
60137	Sk 60A	F20-137, F5-137	
60138	Sk 60A	F20-138, F5-138	Team 60
60140	Sk 60E	F5-140	
60141	Sk 60A	F20-141, F5-141	
60142	Sk 60E	F5-142	
60143	Sk 60E	F5-143	
60144	Sk 60E	F5-144	
60145	Sk 60E	F5-145	
60146	Sk 60E	F5-146	
60147	Sk 60E	F5-147	
60148	Sk 60E	F5-148, FC-38, F5-148	
60149	Sk 60E	F5-149	
60150	Sk 60E	F5-150	

Petter svart, the fifth squadron of F5, undertakes the GTU training course. Its instructors are also the core of the wartime Sk 60 light attack units.

fourth 'squadron' is a training unit, or company (stationskompani) responsible for the wing's conscript ground crew. On 9 January 1995, F5's 1 and 4 Divisionen were combined to become Erik röd, and 2 and 3 Divisionen combined to become Erik blå.

A new student at the GFU begins flying on the Sk 60 straight away – the Sk 61 Bulldog was withdrawn as a primary trainer in 1986. The course lasts for 125 flying hours. Recent air force reorganisations mean that a six-month, 65-flying hour elementary weapons training course (GTU 1) is now also the responsibility of F5.

F16 has three squadrons (Petter blå – JA 37, gul – JA 37, and svart – Sk 60) and maintains the GTU 2 (Grundläggande Taktisk Utbildning/Basic tactical training) course. Twenty-four Sk 60s are allocated to F16's 5 Divisionen, the Sk 60 squadron. The 55-flying hour GTU 2 course encompasses limited air-to-air combat training (including gunnery against towed targets) and air-to-ground tactics. A pilot thus completes 225 hours before transitioning to the Viggen.

GTU 2 has a sub-division – a standing Light Attack Unit (Lätt Attackenheter), equipped with Sk 60B/Cs. This peacetime

training unit would form a major part of Sweden's four wartime light attack units, which would be staffed by its F16 instructors. As F16 has no 'No. 1 Sqn' (Petter röd), the instructors of Petter svart wear red or red and black scarves, an allusion to their 'frontline' status. No such distinction is reserved for students, who wear green scarves as a reflection of their junior status.

Team 60 is the Flygvapnet's six-ship Sk 60A display team, piloted by instructors from F5. Team 60 was established in 1976

Saab 105Ö in Austrian service

The Saab 105Ös were delivered to the three staffeln of the independent Jagdbombergeschwader at Linz/Horsching. The 1st Staffel used yellow codes, the 2nd (flying from Graz/Thalerhof) used red and the 3rd used green. Blue codes were used by the 1st Staffel of the Überwachungsgeschwader (surveillance wing) at Zeltweg.

In mid-1976 the 'Heeresgliederung 75' concept brought about a change in the order of battle. The Jagdbombergeschwader became subordinate to a new Fliegerregiment III, while the 2nd Staffel transferred to the control of the Überwachungs-Geschwader, which had in turn come under the control of Fliegerregiment II. The old green-coded 3rd Staffel took over the second Staffel designation within the Jagdbombergeschwader of Fliegerregiment III and was tasked with reconnaissance, using the newly delivered recce pods. To maximise flexibility, pilots from the 2nd Staffel were also trained in the fighter-bomber role, while pilots from the 1st Staffel trained in reconnaissance tactics and techniques.

This left 1. Staffel/Jagdbombergeschwader of Fliegerregiment III at Linz-Hörsching operating the yellow-coded Saab 105s, and 2. Staffel/Jagdbombergeschwader operating the green-coded Saab 105s at the same base. No. 1. Staffel/Überwachungsgeschwader of Fliegerregiment II remained at Zeltweg with the blue-coded Saab 105s, while 2. Staffel/Überwachungsgeschwader of Fliegerregiment II at Graz-Thalerhof operated the red-coded Saab 105s. The primary roles of the Überwachungsgeschwader's 1st Staffel were training and radiation monitoring, while the 2nd Staffel had a primary role of air-policing. The Überwachungsgeschwader took over the Feuerunterstützung (close support), Luftaufklärung (reconnaissance) and Luftraumüberwachung (airspace surveillance) roles of the Düsenstaffel (jet training squadron) and its motley collection of Vampires, Magisters and J29 Tunnens.

Later, the red squadron merged with the blue, while the yellow squadron merged with the green one. These respectively became the first and second Staffeln of the un-numbered Jagdbombergeschwader at Hörsching, while the Überwachungs-Geschwader lost its assigned Saab 105 units. The Saab 105s were withdrawn from full-time use with the Überwachungs-Geschwader on the arrival of the Saab 35 Drakens. The first Staffel at Horsching continues to provide some eight aircraft to the Überwachungsgeschwader for air policing and support duties, however.

With a major reorganisation taking effect on 1 July, 1995, all remaining 29 Saab 105Ös were directly assigned to the Staffeln of Fliegerregiment III, the Jagdbombergeschwader having disappeared in the re-organisation. Aircraft based at Linz/Hörsching are assigned to a newly-established unit, the Ausbildungs &

Saab 105Ö – Österreichische Luftstreikräfte

Serial	c/n	In service	Notes	Serial	c/n	In service	Notes
A(y)	105401	09.70	Tiger scheme 09.93 / w/o 03.03.95	A(r)	105421	06.71	w/o 05.10.71
B(y)	105402	09.70	1.Sta/JaboG	B(r)	105422	09.71	1.Sta/JaboG
C(y)	105403	09.70	w/o 07.05.75	C(r)	105423	09.71	1.Sta/JaboG
D(y)	105404	09.70	1.Sta/JaboG	D(r)	105424	11.71-	w/o 1993
E(y)	105405	11.70	1.Sta/JaboG	E(r)	105425	11.71	1.Sta/JaboG
F(y)	105406	12.70	1.Sta/JaboG	F(r)	105426	01.72	1.Sta/JaboG
G(y)	105407	12.70	1.Sta/JaboG	G(r)	105427	12.71	1.Sta/JaboG
H(y)	105408	12.70	w/o 16.10.81	H(r)	105428	12.71	1.Sta/JaboG
I(y)	105409	12.70	1.Sta/JaboG – (VIP)	I(r)	105429	01.72	1.Sta/JaboG
J(y)	105410	02.71	1.Sta/JaboG	J(r)	105430	03.72	1.Sta/JaboG
A(g)	105411	02.71	2.Sta/JaboG	A(b)	105431	03.72	2.Sta/JaboG
B(g)	105412	02.71	2.Sta/JaboG	B(b)	105432	05.72	2.Sta/JaboG
C(g)	105413	02.71	w/o 14.05.77	C(b)	105433	05.72	2.Sta/JaboG
D(g)	105414	03.71	2.Sta/JaboG	D(b)	105434	06.72	2.Sta/JaboG – (VIP)
E(g)	105415	03.71	w/o 1.12.76	E(b)	105435	11.72-	2.Sta/JaboG
F(g)	105416	06.71	2.Sta/JaboG	F(b)	105436	06.72	2.Sta/JaboG
G(g)	105417	06.71	2.Sta/JaboG	G(b)	105437	09.72	2.Sta/JaboG
H(g)	105418	06.71	w/o 09.01.77	H(b)	105438	09.72	w/o 11.10.77
I(g)	105419	06.71	w/o 06.08.81	I(b)	105439	11.72	2.Sta/JaboG
J(g)	105420	06.71	w/o 06.04.88	J(b)	105440	11.72	2.Sta/JaboG

Einsätzstaffel/Duse. In the new Army Organisation (Heeresgliederung-NEU) the Saab 105Ö is assigned to the advanced training and air-policing roles only, and the reconnaissance and close air support functions have formally been abandoned.

Two Austrian aircraft (currently 105409 yellow I and 105434 blue D) are permanently maintained in four-seat VIP configuration. Other aircraft can be

converted to the same standard using the air force's two conversion kits.

105407, 105423, 105425, 105426, 105427, 105428, 105429, 105431, 105432, 105433, 105434, 105435, 105435, 105436, 105437 and 105439 all served with the 'Silver Birds' or 'Karo As' team. Of the write-offs, 105413, 105415, 105418, 105438 were all noted in the scrap compound at Linz in 1978.

In addition to being advanced 105XT standard aircraft, Austria's Saab 105Ös also possess a unique photo-reconnaissance pod, built around an array of Vinten cameras. The 105Ös replaced J29F Tunnens, which could be fitted with a ventral camera pack, at the expense of two of their internal cannons.

Mitsubishi F-1

Ship-killing Samurai

A derivative of the T-2 trainer described in *World Air Power Journal* Volume 18, the Mitsubishi F-1 has performed the vital anti-shipping strike role since the late 1970s, and the three remaining squadrons seem likely to continue in service until the introduction of the Mitsubishi FS-X at the turn of the century. The aircraft's role is officially described as being 'anti-landing craft', a suitably defensive term for what is, in reality, an impressive offensive maritime strike capability.

A populous and highly industrialised island nation with dwindling natural resources, Japan is utterly dependent on foreign trade. Its raw materials generally come from overseas, while markets are also chiefly found outside Japan. All of this brings with it a heavy reliance on the sea, and defence of the sea lanes is accorded a high priority. This stimulated Japan's growth as a major naval power during the early part of the 20th century. Similarly, any foe of Japan would have to invade the islands which constitute the nation by sea or by air. Auto-nomous airborne invasion is impractical, even for a superpower, and defence against seaborne invasion has thus always been viewed as being absolutely crucial to the nation's security.

Nevertheless, for many years Japan was unable to equip its armed forces with equipment capable of defending against invasion or against interdiction of the sea lanes. After Japan lost World War II, the victorious Allies ensured that the defeated nation remained militarily weak. Allied strictures, combined with the humiliation of the old military class and the growth of a culture of anti-militarism, made it inevitable that Japan's new armed forces would be overtly defensive in nature. The three arms were even named as 'Self-Defence Forces' and were equipped with defensive weapons and systems. Thus the anti-shipping role is known as the 'anti-landing craft' role, and fighter-bombers are known as support fighters.

For many years, the Japan's Air Self-Defence

Laden with four 500-lb bombs, and closely tailed by an F-104J chase aircraft, the first FS-T2 Kai is seen on approach to Gifu during early development trials. The two FS-T2 Kai aircraft were converted on the line from T-2 airframes.

Left: An F-1 from the 3rd Hikotai blasts off from Misawa, both Adours in full afterburner. The F-1 is seldom flown without at least one auxiliary fuel tank, as its range on internal fuel is rather limited.

Above: The first prototype Mitsubishi F-1 was rolled out in a smart operational camouflage colour scheme. This, and the absence of a rear cockpit, set it apart from the grey and Dayglo FS-T2 Kais.

requirement for a close-support and fighter/fighter-bomber aircraft, with minimal modification. This part of the JASDF requirement specified that the aircraft should be able to fly an air combat profile (designated Profile I) and a ground support profile (Profile II). The specified air combat profile consisted of a scramble take-off with a climb to 11000 m (36,100 ft) in full afterburner, flying 150-200 km (93-125 miles) at Mach 0.9 before accelerating to Mach 1.4 prior to five minutes of combat and a return to base with 15-20 minutes of reserve fuel. No external fuel was to be carried.

The specified air-to-ground profile included a high-level transit for 350 km (217 miles), with a 150- to 200-km (93- to 125-mile) low-level dash to the target and five minutes in the target, before a lo-hi transit home. Reserves of 15 minutes were again specified, and the weapon load was set at two 750-lb bombs, with external fuel.

Indigenous or imported?

It was never inevitable that an indigenous aircraft would be designed to meet the twin requirements, not least because any such aircraft

would have a tiny production run, and would thus have a high unit price. Some believed that the T-38 (or F-5B) should be adopted for the training role, with F-5As as the close support fighter type. Others within the Japanese defence establishment favoured procurement of the Anglo-French SEPECAT Jaguar.

The Jaguar was intensively evaluated, and the JASDF was extremely impressed by the aircraft, which seemed to offer all the performance, handling and operational characteristics required for the Japanese advanced trainer and fighter-bomber aircraft. Simply buying Jaguar off-the-shelf was not acceptable to Japan, however, since the trainer and fighter-bomber programmes were seen as a means by which Japan's aerospace industry could gain valuable experience. Accordingly, Japan conducted negotiations with the British and

The first prototype F-1 went on to serve with front-line squadrons after the test programme was complete. It is seen here while serving with the 3rd Kokudan at Misawa during 1992, where it continues in active use today. For a prototype to see such extensive operational use is extremely unusual.

Force's front-line strength consisted entirely of North American F-86F Sabres. New aircraft types introduced subsequently were similarly dedicated fighter-interceptors, from the Lockheed F-104 Starfighter to the McDonnell Douglas F-4 Phantom. A handful of the Sabres were assigned to the tactical fighter role (despite their unsuitability and lack of dedicated air-to-ground weaponry), and the F-4s and F-104s assumed a secondary ground attack role, but by the 1970s the lack of a dedicated ground attack and anti-ship aircraft was becoming embarrassingly apparent.

Replacing the F-86

Japan also lacked a high-performance advanced trainer, and with the advent of the F-104J this was felt to be the single most important priority for the JASDF, since existing types were felt to be inadequate to prepare pilots for the Starfighter. It was felt that such a trainer aircraft (the so-called T-X) would need to have very similar performance and handling characteristics to any front-line fighter-bomber, and it was noted that the F-86F had already performed quite successfully in both the advanced training and ground attack roles, proving that one aircraft could be both trainer and fighter-bomber. It was thus soon decided that whichever aircraft was adopted to meet the advanced trainer requirement would have to be capable of mounting the M61A1 cannon, and would also have to meet the SF-X

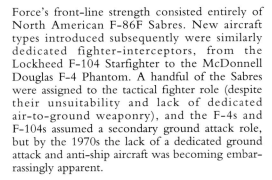

French to build Jaguar under licence, and these negotiations only stalled because Japan was reluctant to pay more than 5 per cent royalties (16 per cent then being normal). Japan's enthusiasm for the Jaguar should not be underestimated, and had a major influence on the configuration and detail design of its indigenous alternative.

An indigenous solution

The decision to proceed with an entirely indigenous trainer was taken in November 1966,

after intensive lobbying by Japanese aircraft manufacturers, and in September 1967 Mitsubishi was awarded a development contract for the T-2, with a basic design contract following on 30 March 1968. The aircraft was selected in preference to designs by Kawasaki and Fuji, Kawasaki withdrawing from the competition before selection. Announcement of the T-2's selection did not stop the JASDF requesting £28.75 million to purchase 31 interim advanced trainers, ostensibly because the existing F-86 and T-33 fleets could not last until the T-2 was introduced. These could have been F-5s or T-38s, while SEPECAT had leapt in with an offer to lease 30 two-seat Jaguars to the JASDF during the period 1971-74. Both Northrop and SEPECAT hoped that a lease of their aircraft, combined with any development problems suffered by the T-2, might prompt a change of heart. In fact, a structural survey revealed that both the F-86 and the T-33 had plenty of fatigue life remaining, and the question

of an interim aircraft faded.

The T-2 prototype made its maiden flight on 20 July 1971 and became Japan's first supersonic trainer on its 30th flight. The new trainer was developed from the start with a view to providing the basis for a fighter-bomber, and Mitsubishi continued feasibility studies of a fighter-bomber version of the aircraft. The Ministry of Finance had other ideas, however, and pushed for purchase of the F-5A in the fighter-bomber role. They took this stand because the small number of fighter-bombers required would impose an inevitably high unit price, and because they felt that only one new indigenous aircraft development programme could be funded.

Political, technological and industrial considerations favoured indigenous development of the P-XL, Kawasaki's P-2J replacement, which promised to keep the Japanese electronics industry at the cutting edge of avionics development. The project was also a high-risk one, however,

Surrounded by ground support equipment, Mitsubishi F-1s of the 3rd Hikotai are prepared for flight on a rain-swept Misawa ramp. The aircraft carry underwing fuel tanks and a centreline practice bomb dispenser.

and Lockheed's P-3 Orion was an obvious alternative. By contrast, there were fewer viable alternatives to the indigenous fighter-bomber whose development (as a version of the T-2) promised to be uncomplicated, to save precious foreign currency, provide high-tech airframe and aerodynamic design experience, and to significantly boost the capabilities of the Japanese aircraft industry. Moreover, there was even some hope that a trainer or fighter-bomber could be exported in small numbers. There were also those who wanted indigenous participation in a very front-line aircraft, the aircraft industry having spent the post-war period trying to forget its past as a pillar of the discredited Imperial military industrial complex. National pride required that companies like Mitsubishi should again build a genuinely military aircraft of its own design.

Mitsubishi was certainly ready for the challenge, having licence-built some 300 F-86F Sabres and 230 F-104 Starfighters. The company had also built Sikorsky SH-3 Sea King helicopters under licence. Additionally, Mitsubishi had played a part in the design of the upgrade which saw a handful of JASDF Sabres converted to RF-86F configuration.

Teaming for the trainer

Moreover, Mitsubishi was not having to 'go it alone' on the new trainer, but was co-operating closely with a number of equally experienced aerospace companies. After the 5 September 1967 adoption of the Mitsubishi design, on 15 October, a 75-man design/planning team was formed under the name ASTET (Advanced Supersonic Trainer Engineering Team) with engineers from Mitsubishi, Fuji, Kawasaki, Shin Meiwa and Nihon Hikoki. The number of engineers eventually reached 175, and the team continued until March 1969 when it disbanded, its task complete. Workshare on the T-2 (and subsequently on the F-1) was divided between the ASTET members, with Mitsubishi taking the lion's share with 70 per cent including final assembly, flight test and manufacture of front and centre fuselage sections. Fuji took 27 per cent with tail, rear fuselage and wing manufacture,

A head-on view of the F-1 reveals many similarities with the Anglo-French Jaguar. The square-section high-set intakes, ventral airbrakes, and anhedral tailplane are all strongly reminiscent of the Jaguar. The SEPECAT Jaguar was almost selected by the JASDF, before the indigenous T-2/F-1 programme was finalised.

while the remaining 3 per cent was divided between Nihon Hikoki and Shin Meiwa, including manufacture of pylons, launch rails and drop tanks. Kawasaki received no workshare, and severed its links with the project in October 1968 to concentrate on licence-manufacture of the F-4EJ Kai. Generally, the same team that designed the T-2 was the same team that would build the F-1.

Eventually Prime Minister Kakuei Tanaka cancelled the P-XL programme, and, in 1972, the government decided to fund the development of an indigenous fighter-bomber based on the T-2. The Northrop F-5 was finally rejected.

Prototype development

A prototype development contract was placed by the Defense Agency in 1973. Mitsubishi was commissioned to take the sixth and seventh T-2s from the Komaki production line, and to rebuild them as single-seat fighter-bombers. These two aircraft were initially known as Special Spec T-2s, and then as T-2(FS)s, before taking the designation FS-T2 Kai.

Changes from the T-2 were kept to an

absolute minimum, in order to minimise cost and delay. The fuselage shape and cross section was maintained, even though drag could have been reduced by getting rid of the bulge where the rear cockpit used to be. Instead the rear cockpit was adapted as an avionics bay, and was covered by a simple access hatch, which followed the shape of the old rear canopy. Because virtually all of the changes were internal, this meant that T-2 flight test results remained largely valid, and FS-T2 Kai flight tests could be limited mainly to exploring the effect of carrying various external stores, and to proving the new avionics and systems.

The seventh T-2 (59-5107) was the first of the two aircraft to fly in single-seat form, on 3 June 1975, with the sixth (59-5106) following on 7 June. The reason for this was that with the rainy season approaching and the first of the two pro-

Mitsubishi F-1

A 3rd Hikotai Mitsubishi F-1 tucks up its undercarriage after take off. The main units retract forwards, while the nose gear retracts to the rear.

totypes laid up for the installation of test equipment, the then-uninstrumented second prototype could be made ready for flight first. The two aircraft were quickly delivered to the Air Proving Wing where they undertook a 230-flying hour flight test programme which lasted until March 1976. 59-5106 was used primarily for performance and aerodynamic tests, with 59-5107 performing tests related to the fire control system. This progressed very smoothly, which was hardly surprising, since aerodynamic, engine, hydraulic, electrical and many other potential problems had been shaken down during the flight testing of the all but identical T-2 prototypes. During initial testing of the T-2, longitudinal stability was found to be inadequate in the approach configuration, but this was solved by reprofiling the wing leading-edge root extension (LERX) and the wing fences. Further fixes were applied to the underwing pylons, to improve stability when

A quartet of 6th Hikotai Mitsubishi F-1s lines up at Tsuiki. The nearest aircraft has its brake chute compartment door open, probably indicating that the aircraft have just returned from a practice bombing mission.

stores were being carried. The tendency to yaw caused by the offset location of the nose gear was solved at the same time, by adding a vertically-mounted aerofoil surface on the oleo.

Production orders

At the conclusion of the test programme the JASDF placed its first order for 18 of the new aircraft. The JASDF had wanted to place an initial order for 50 aircraft in Fiscal Year 1976, but was unable to do so because of the deteriorating financial situation, and was forced to 'drip-feed' the order over several fiscal years. A large single order would have allowed more rapid re-equipment of the three-squadron fighter-bomber force, whereas limiting the original order to only 18 aircraft delayed formation of the second and third units by one and two years, respectively. The initial 18 aircraft order preceded the 12 November 1976 award of a type certificate, under the production designation F-1. Director General of the Defence Agency, Mr M. Sakata, formally authorised use of the F-1 designation and formally cleared the aircraft for use by operational units.

The two FS-T2 Kai aircraft appeared externally identical to the T-2 prototypes which preceded them. The aircraft even retained clear canopies over the avionics bays which replaced

the rear cockpits, although inside these were metal covers. They retained the same gloss grey colour scheme, with the same Dayglo fins, noses and wingtips. The only external distinguishing feature of the FS-T2 Kai was the 3-in (7.5-cm) diameter tubular RHAWS antenna fairing which capped the tailfin. The only other difference between the FS-T2 Kai prototypes and the T-2 development aircraft thundering in and out of Mitsubishi's test airfield at Komaki lay in the stores they carried. These included ASM-1 anti-ship missiles, bombs, rockets, ECM pods and even dart target-towing pods, usually in conjunction with test camera pods underwing or on the centreline.

Production prototype

The first true F-1 prototype, 70-8201, was rolled out at Komaki on 25 February 1977, and made its maiden flight on 16 June 1977 in the capable hands of Kenshiro Endo. This was almost nine years after the first flight of the Jaguar, which made the Japanese aircraft seem quite

up-to-the-minute. 70-8201 underwent brief manufacturer's flight trials, before being handed over to the JASDF on 16 September 1977. This aircraft later entered squadron service, and remains in front-line service today.

First squadron

The first JASDF unit to convert to the Mitsubishi F-1 was the 3rd Hikotai at Misawa, which began re-equipment in September 1977, transferring to the control of the 3rd Kokudan on 1 March 1978 when conversion was complete. The 3rd Kokudan's second squadron, the 8th Hikotai, began conversion to the F-1 on 30 June 1979, and when this was complete the 6th Hikotai began transitioning to the new type, on 11 March 1980. This was the sole F-1 unit assigned to the 8th Kokudan at Tsuiki. Squadrons originally formed with 18 aircraft each (plus attrition replacements), but this was felt to be too small a number, and the JASDF pushed for

squadrons with 25 aircraft each. This prompted a move by the Japanese defence ministry to cut the number of squadrons to two, and the JASDF rapidly decided that three 18-aircraft squadrons were ideal, or at least better than two 25-aircraft units.

The F-1 (like the T-2) has been no stranger to political intervention and controversy and has faced cancellation, termination and reductions on several occasions. Perhaps the most difficult period came during the 1970s, when rampaging inflation remorselessly drove up the aircraft's unit cost, and while Western arms salesmen tried to tempt the JASDF with low prices for their aircraft. Once the aircraft was in production new

*Right and below: **This F-1 wears an interesting motif, a bat-winged nude female devil, on all fours, holding a huge weapon in her hand. Such individual markings are by no means uncommon on JASDF aircraft, though they are often fairly short-lived.***

A 6th Hikotai F-1 lands at Tsuiki, while an E-2C waits at the holding point. The F-1 does not have flap blowing, so take off and landing speeds are relatively fast. This aircraft's camouflage scheme has faded markedly.

problems emerged; Mitsubishi reacted angrily to the slow supply of parts from overseas sub-contractors, although Ferranti was singled out as being an honourable exception.

Mitsubishi F-1

Left and above: This ASM-1 armed F-1 of the 8th Hikotai also carries a centreline fuel tank decorated with the unit's previous tail markings, commemorating the squadron's 30th anniversary, and recording its service at Komatsu (1961-1964), Iwakuni (1964-67), Komaki (1967-1978) and Misawa. The indigenously developed ASM-1 is broadly comparable to the German Kormoran and French Exocet anti-ship missiles, with similar range and warhead characteristics. A replacement for this weapon, the ASM-2, is at an advanced stage of development. It is believed that the F-1s would usually employ third-party targeting (perhaps provided by JMSDF Orions) for their ASM-1 missiles, since their own onboard radar lacks range.

The F-1 retained the same engine as the T-2. Early in the history of the project it had been decided that the aircraft would be powered by an indigenously-built version of a foreign engine. The General Electric GE-1 was a strong contender despite existing only on paper, but the Rolls-Royce Adour was the JASDF's favourite. It can hardly have been coincidence that the Adour was the powerplant of the SEPECAT Jaguar that the JASDF coveted.

The Anglo-French Rolls-Royce/Turboméca RB.172/T-26-40 Adour was thus built under licence by Ishikawajima Harima Heavy Industries as the TF40-IHI-801A, following its 15 February 1968 selection for the T-2. The Adour has proved an extraordinarily good engine, with superb reliability and low specific fuel consumption. Unfortunately, early versions of the engine were lacking in thrust, and initial operators of the Jaguar found that the aircraft was underpowered. Export Jaguars were fitted with later, more powerful Adours, while the engines of RAF Jaguars have since been considerably uprated. The F-1, on the other hand, soldiers on with low-powered early series Adour 801s, broadly equivalent to the Adour 101s that were initially fitted to RAF

Jaguars but quickly replaced. The first prototypes were powered by British-built RB.172/T26-40 Adours – these engines proved reliable enough, but early Japanese-built engines suffered from many irritating teething troubles, including poor throttle response. Huge engine doors below the fuselage give superb unimpeded access to the powerplants.

Low drag

When the T-2 had been designed, low drag, simple structure and low weight had been nominated as primary design objectives. As a result the F-1 is extremely lightly built. The fuselage, for example, is 25 per cent lighter than that of the smaller F-104J Starfighter. The aircraft's empty weight is similar to that of the T-2 and increased maximum take-off weights are achieved without making major structural changes and without changing materials .

The fuselage is entirely conventional, of semi-monocoque design and square section. It is area-ruled to minimise drag and is built in four parts. The bulk of the fuselage (66 per cent) is of high-tensile aluminium alloy, with magnesium alloy and other alloys for some parts (7 per cent),

including the canopy frames, and titanium alloys elsewhere (9 per cent). Little use is made of composites or plastics.

Behind the radome, tipped by a fixed pitot, is the forward avionics compartment, and behind that is the oxygen compartment, with a built-in LOX converter. Below the pressurised cockpit are the JM61 cannon bay and the nosewheel well.

The cockpit is covered by an upward-opening stretched acrylic canopy and accommodates a locally licensed version of the Weber ES-7 ejection seat, with locally designed canopy penetrators. The seat is fitted with an LRU-3P dinghy, an oxygen cylinder, survival equipment and a chaff dispenser. The chaff dispenser is used to generate a radar trace to enable the ejected survivor to be pinpointed.

The ammunition tank is behind the avionics bay which occupies what was the T-2's rear cockpit, above the No. 1 fuel tank. The centre fuselage contains tanks No. 2 to No. 5, between the intakes and intake ducts. Downward-opening air brakes, hinged on the leading edge, are fitted on the lower fuselage, immediately behind the main undercarriage bays.

The rear fuselage houses the No. 6 tank, behind which is the No. 7 tank, and below that are the two side-by-side Adours, which are separated by a titanium keel. The keel extends aft behind the jetpipes, and serves as the mounting point for the heat-resistant Iconnel airfield arrester hook.

The single-piece Delta-type wing incorporates 9° of anhedral, and the leading edge incorporates 42° 29' of sweepback, plus a dogtooth disconti-

The Mitsubishi F-1 remains Japan's only indigenous front-line aircraft. The Mitsubishi FS-X which is scheduled to replace it is arguably less of an indigenous design, being little more than a Japanese sub-variant of the F-16 Fighting Falcon. It is becoming increasingly more difficult for single nations to develop advanced combat aircraft without international partners, and aircraft like the F-1 mark the end of an era.

nuity which helps keep flow attached by generating a powerful vortex. Despite the dogtooth which divides it in two, the leading-edge flap functions as a single unit. Full-span flaps leave no room for ailerons, and roll control is therefore achieved using spoilers. Spoilers alone are not an ideal method of controlling an aircraft at all speeds, and are difficult to design. Mitsubishi had, however, effectively prototyped the spoiler control system on its MU-2 business aircraft. The trailing-edge flaps and spoilers are each in two sections. The spoilers are linked, but separately actuated. A small fence is fitted at the root, immediately inboard from the LERX. The wing forms a box-beam structure, with front and rear spars at 15 per cent and 56 per cent chord.

The horizontal tail is set slightly lower than the wing and incorporates 15° of anhedral. The high root keeps it clear of the jet efflux, while the low tip remains clear of the wing downwash. The inboard part of the tail is constructed of titanium and steel alloys, while simple aluminum alloys are used further outboard. Titanium leaf is applied to the leading edge when rockets are carried, to protect the leading edge from rocket blast. The all-flying tailplane has leading-edge sweep of 42° 32' and moves through +9° and -20.5°. The tailplane is built around a box-beam main spar, with separate leading and trailing edges and tips.

Tail unit

The vertical tail carries the tubular antenna fairing for the RHAWS, while the J/ARC-51 UHF aerial is inside the fin cap. The rudder is of aluminium alloy and honeycomb construction, and is attached to the fin by three hinges, and is actuated by a single hydraulic servo. The bottom edge of the rudder incorporates a cut-out to avoid fouling the tailplane.

Simplicity is assured by the elimination of vari-

The 8th Hikotai was the second unit to re-equip with the Mitsubishi F-1, and is the second of two F-1 squadrons assigned to the 3rd Kokudan at Misawa. The squadron's black panther badge was adopted in 1983.

ous systems. Thus, despite the aircraft's tiny, highly-loaded wing, it has no boundary layer control or flap blowing system. Simple flaps on the leading and trailing edges are the only high-lift devices.

Although of remarkably heavy-duty construction, the undercarriage is also simple and relative-

This Mitsubishi F-1 carries a target-towing winch under its starboard wing, though the dart target which would be attached to this is missing. The F-1 is seldom used as a target tug, except when F-1 squadrons practise air-to-air gunnery in the run-up to competitions and annual armament practice camps.

Mitsubishi F-1

ly light. Designed and built by the Sumitomo Precision Metal Co. Ltd, the gear is hydraulically actuated, with an emergency compressed air system to lower the undercarriage in the event of a hydraulic failure. The main landing gear retracts in 6.5 seconds. The undercarriage is shod with tyres similar to those used by the F-104J, 18 x 5-in (12.7-cm) on the nosewheel and 26 x 6.75-in (17.2-cm) on the mainwheels. The nosewheel is always kept at 161 psi, while the mainwheels can be inflated to between 214 and 992 psi, depending on aircraft weight.

Apart from the installation of a J/APR-3 RHAWS with its distinctive fin-mounted antenna, the F-1 has a new J/ASN-1 inertial navigation system, a J/ASQ-1 weapons release computer and a J/A24G-3 air data computer. With

improved navigation equipment came J/APN-44 radar altimeter antennas set into the lower part of the nose, with the total temperature probe relocated to the starboard side of the nose. The passive radar warning receiver system was developed and built by Tokyo Precision Instrument, and soon proved extremely successful. Most other avionics items are shared with the T-2 and include J/ARC-51 UHF radio, J/ARN-53 TACAN, AN/ARA-50 UHF DF, J/APX-101 SIF/IFF (Selective Identification Feature/ Identification Friend or Foe).

Armament

Like the last 62 two-seat trainers (the T-2(K) Koki aircraft), the F-1 is fitted with an internal JM61A1 20-mm Vulcan cannon. While the T-2(K) carries 600 rounds of ammunition for this formidable weapon, the F-1 carries 750 rounds. External ordnance can include the AIM-9 Sidewinder in its E or L versions. The AIM-9E was built locally as the AAM-1. When carried, Sidewinders can be fitted to LAU-7/AA launchers on the wingtips and under the outboard underwing pylons. External weapons and stores are carried on pylons mounted on five hardpoints, two under each wing and one under the

centreline. These can each carry a single JM117 750-lb bomb, or a range of other stores, including 250-lb, 500-lb or 750-lb slick and retarded bombs or cluster bombs. Up to 12 500-lb bombs can be used by using Double Ejector Racks or Four Ejector Racks (FERs). Other common F-1 weapons include the JLAU-3A rocket pod, containing 19 70-mm rockets, or the RL-4 or RL-7 pod, which respectively contain four 125-mm or seven 70-mm rockets.

In the anti-shipping role the primary weapon is the Type 80 ASM-1, which also equips JMSDF P-3 Orions. Developed in a programme which began in 1973, by Fuji, Kawasaki, Mitsubishi and the Defence Agency's Technical Research and Development Institute, the ASM-1 is powered by a solid fuel rocket motor, and has a range of about 50 km (30 miles). It employs inertial guidance and has active radar terminal homing. The missile has four fixed Delta wings and four tail-mounted triangular control surfaces. The semi-armour piercing warhead weighs 150 kg (330 lb). Prototype missiles were flown in 1977, and the weapon was evaluated by the JASDF during 1981 and 1982, entering service in 1983. The ASM-1 is due to be replaced by the Type 88 ASM-2 later in the decade, and this new missile may enter service with the F-1 fleet.

A recce F-1?

There were plans for a dedicated reconnaissance version of the F-1/T-2 family, provisionally designated RT-2. This aircraft would have replaced the RF-86F, but in the end the tried and tested RF-4EJ was selected instead. The RT-2 did not reach prototype stage, and details of its reconnaissance fit have not emerged.

Few detailed assessments of the F-1 have been made or published. Although Japan has a thriving aviation publishing industry, the majority of books and magazines produced are aimed at the enthusiast and modeller, and offer more detail on history and colour schemes than on capabilities and handling. Few non-Japanese pilots have ever flown the F-1, and their opinions of the aircraft

A pair of 3rd Hikotai F-1s taxis out for a training mission. The aircraft can seldom be seen carrying weapons, most training sorties being conducted using external fuel tanks only. The aircraft has no inflight-refuelling capability.

have not been published. Interviews with Japanese F-1 pilots have not been terribly revealing or informative, beyond suggesting that the cockpit is 'optimised for the average Japanese physique' and that the aircraft has low stick forces and 'mild' control sensitivity. Published specifications would seem to suggest that by today's standards the aircraft is both underpowered and underfuelled, with a barely adequate radius of action, except with external fuel, and with a tiny warload. Early reports also suggested that radar performance was not initially up to the mark, and it seems unlikely that much improvement was made thereafter.

The problems which rendered the T-2 trainer ineffective in the adversary role are shared by the F-1. The type's lack of thrust and agility make it easy meat for more powerful fighters, while vicious departure characteristics demand skilled handling and impose an unnecessarily high cockpit workload. The F-1 did mark a major improvement over the F-86F Sabre which it replaced, however, and kept Mitsubishi in the fighter-building business, providing invaluable experience for later programmes.

Today the Mitsubishi F-1 is looking increasingly long in the tooth, and its replacement, the FS-X, remains a long way from service. There have been reports that one of the JASDF's Phantom squadrons is to be retasked in the maritime strike role, and that this unit, together with two F-1 squadrons, will serve on until the FS-X enters service. The low-houred surviving F-1s

All three F-1 squadrons retain a number of T-2 dual-control trainers. Nowadays these are painted in F-1-style camouflage and wear squadron markings, and are thus difficult to tell apart from the single-seat operational aircraft.

will undergo a SLEP to extend their lives into the next century. Despite many weaknesses and encroaching obsolescence, the F-1 remains extremely popular with its pilots, who are proud to fly what is still Japan's only indigenous front-line fighter. **Jon Lake.**

Above: The F-1 is a large and heavy aircraft for a single-seater, and the use of a braking chute is routine. The aircraft does not have entirely vice-free handling characteristics, and this, together with the aircraft's demanding low-level role, means that it is not considered a suitable posting for any but the best first-tour pilots.

Mitsubishi F-1
8 Hikotai (Eighth Squadron)
3 Kokudan (Third Air Wing)
Hokubu Koku Homentai (Northern Air Defence Command)
Misawa Air Base
Honshu, Japan

Particularly in recent years, JASDF Mitsubishi F-1s have worn colourful markings for participation in exercises and competitions. This aircraft was one of nine which received new rudder decorations for the 1994 Senkyo (ACM/gunnery competition) at Chitose. This aircraft was actually the reserve, and had a painting of Tenyo, a celestial woman, applied to the starboard side of the rudder, with the legend 'Come and see' to port. The other eight team aircraft had representations of characters from Nanso Satomi Hakken Den (the Legend of the Eight Dogs of Nanso) by Bakin Takizawa. In this Edo period (1603-1863) novel a maiden was impregnated by the spirit of a dog, giving birth to eight warriors who then fought against the evil Satomi family.

Mitsubishi F-1 colours

With rare exceptions, the bulk of the F-1 fleet has continued to wear the same three-tone camouflage colour scheme in which it was delivered, with two shades of green in a disruptive pattern over a tan base colour. This mimicked the USAF's Vietnam-era camouflage, though the shades used were different, with one of the greens being much brighter. This camouflage pattern extends down to the lower corners of the fuselage, with light grey on the undersides. National insignia is carried in the form of white-outlined Hinomaru on the sides of the forward fuselage (behind the cockpit) and above and below each wing. The six-digit serial is carried quite high on each side of the tailfin, and the 'last three' are repeated as a tactical code on the sides of the nose, approximately centred on the canopy arch.

During the late 1980s, a number of alternative colour schemes were applied to the F-1s, mostly designed to reduce the aircraft's conspicuity when flying at low level over the sea. It is not known which aircraft were involved in the trials, nor are the exact dates of repainting the aircraft, or of restoring them to standard colours.

Several different aircraft from each Hikotai were observed in non-standard colour schemes between 1986 and 1990, however, and these aircraft are detailed below.

The first solution to the problem was

to replace the base tan colour with a blue-grey shade, and this was tried on 10-8259 of the 3rd Hikotai (seen in mid-1990), with a light blue grey replacing the tan, and with the new top surface camouflage colour continuing to wrap-around the lower surfaces. The unit badge was toned down by light overpainting but the nose codes were made more conspicuous, being thinly outlined in yellow. 70-8206 and 80-8215 of the 6 Hikotai (seen in August 1989) were similarly painted, but with a much darker blue replacing tan, and again with upper surface camouflage wrapping around to cover the lower surfaces. Also in the 6th Hikotai, 80-8211 and 90-8228 replaced tan with a more neutral light grey. 90-8228 had its Hinomaru reduced in size, toned down, and relocated to the upper part of the rear fuselage, adjacent to the fin leading edge, also losing its squadron badge, serial number and nose code. The simple code 28 was applied on the side of the intake, below the wing leading edge, and a miniature version of the badge was applied slightly further forward. Both of these markings were in grey on the green camouflage colour. 80-8210 of the 8th Hikotai (seen in September 1986) had both shades of green replaced by olive drab, and tan replaced by a medium sea grey. It wore toned-down Hinomaru of reduced size, with a thin white outline. Codes, serial and unit badge were almost obliterated by overpainting.

A number of F-1s were repainted altogether, with all the original upper surface camouflage colours being replaced. 70-8209 of the 3rd Hikotai (seen in January 1987) had thinly applied

dark sea grey topsides, with the old green areas showing through as a darker shade. The Hikotai badge was applied in black outline form, while the Hinomaru were of reduced size, and toned down, with no white outline. The same aircraft transferred to the 6th Hikotai, where (by mid-1990) it was again repainted, with full-size, full-colour markings and squadron badge, but with a new disruptive camouflage of dark sea grey 'splodges' over a medium sea grey base colour.

Sister aircraft 80-8211 (seen in August 1989), had a similar colour scheme, but with the darker grey applied in bands which more closely followed the green areas of the original green/tan camouflage. Another 6th Hikotai aircraft wore a similar camouflage pattern, but with medium grey forming the darker colour, and with a pale blue grey base. This aircraft had a tiny outline squadron badge in medium grey on the fin, and a small Hinomaru low on the rear fuselage, centred on the red warning stripe. Serials and codes were not carried, with only the two-digit code '19' on the side of the intake, applied very small. One aircraft

Overall shades of matt sea grey and toned-down insignia made this one of the least conspicuous F-1s in the maritime environment.

from the 8th Hikotai (80-8212, seen in September 1986) tested an even more ambitious maritime colour scheme, with three shades of blue grey. This comprised a very dark shade covering the top surfaces, with a marginally lighter shade on the fuselage sides and sides of the tailfin, plus a significantly lighter shade on the undersides. The aircraft replaced its national insignia with simple black rings, and the panther unit badge was sketched in in black lines. Codes and serial were reduced in size. None of these colour schemes was adopted for the fleet, and today most F-1s wear their original camouflage. Individual markings are frequently applied for exercises.

The camouflage worn by this F-1 is very similar to the standard overland colour scheme, but with a light sea grey replacing the usual tan. The aircraft served with the third Hikotai.

Two-tone blue grey in a disruptive pattern decorated this 6th Hikotai F-1. High-visibility orange flying suits are favoured by the JASDF.

60

Serial system
Like all JASDF aircraft, the F-1 uses a two-digit prefix indicating delivery date and aircraft type, followed by a four-digit serial with a role designator and a three-digit sequential build number (which commences at 101). Thus the aircraft shown here (60-8272) was delivered in 1986, is a Mitsubishi F-1 (0) used in the tactical fighter role (8) and was the 171st aircraft delivered (101 + 171 = 272). The last three digits are repeated on the sides of the nose, as a tactical code. Aircraft do not have a separate, discrete code or numerical identity within a squadron.

Wing
The one-piece wing is built by Fuji, and is fitted with leading-edge flaps and powerful, full-span flaps on the trailing edge. Plans to fit a flap-blowing and boundary layer control system were abandoned in order to save weight and to reduce cost and complexity. The aircraft uses spoilers and differential taileron movement for roll control, and is therefore not fitted with ailerons. The wing has no integral fuel tanks, all fuel being carried in fuselage tanks, or in external auxiliary drop tanks.

Hardpoints
The F-1 has a total of five external hardpoints, one on the centreline and four under the wing. These are usually fitted with Nippi-made pylons. These can each carry a single JM117 750-lb bomb (or equivalent). A total of 12 500-lb bombs may be carried by using DER (Double Ejector Racks, equivalent to USAF TERs and capable of carrying two bombs) or FER (Four Ejector Racks, equivalent to USAF MERs and capable of carrying four bombs) racks on the underwing pylons. The inboard underwing and centreline pylons can be used for the carriage of up to three 220-US gal (821-litre) fuel tanks. These are usually fitted with asymmetric cruciform tailfins to improve jettison characteristics, forcing the tank's nose downwards and preventing the falling tank from flying up into the aircraft. The ASM-1 anti-ship missiles which now form the F-1's primary weapon can be carried on the inboard underwing pylons only. Other weapons commonly carried by the F-1 include the RL-4 rocket pod, containing four 127-mm rockets.

Fin markings

This aircraft wore the kanji Kenzan, meaning 'Come and see' on the port side of the rudder, and the figure of Tenyo, a celestial woman, on the starboard side. Although she is drawn in the traditional style, she is carrying an AIM-9 Sidewinder in her left hand! The aircraft was the reserve for the 8th Hikotai team participating in the 1994 ACM/gunnery competition at Chitose. The other aircraft were decorated with similarly stark black and white figures, armed with more traditional Samurai weapons. The squadron's permanent badge is a black panther, outlined thinly in yellow.

Tail unit

The tailplane is a one-piece cantilever unit, with titanium inboard leading edges and aluminium honeycomb trailing edges. The fin has a single-piece hydraulically-actuated rudder. The entire tail unit, like the wing, was built by Fuji at Utsonomiya. Underwing stores pylons were produced by Nippi, and the undercarriage was built by the Sumitomi Precision Metal Company. The aircraft were assembled on the same line as the T-2s, by Mitsubishi at Komaki.

Canopy

The cockpit is covered by a single-piece upward-opening clamshell canopy. This does not incorporate MDC and so must be jettisoned or ejected through.

HUD

The Mitsubishi F-1 is fitted with a simple narrow field-of-view HUD produced by Mitsubishi Electric under a licence agreement granted by Thomson-CSF. The cockpit itself is of typical early 1970s vintage, with conventional analog instruments, and the lower part of the panel is dominated by the hooded circular radar display.

Radio altimeter

The F-1 has a J/APN-44 radio altimeter, with twin antennas below the nose. This allows more accurate height keeping at low level. Surprisingly, the aircraft does not have any terrain-following or terrain-avoidance capability. A total temperature probe is fitted below the port side of the cockpit. Neither of these equipment items is fitted to the T-2 trainer.

Wing

The cantilever wing incorporates 9° of anhedral and uses a modified NACA 65 series profile, with a thickness/chord ratio of 4.66 per cent. The wing is constructed of 7075 aluminium alloy, and is built around a multi-spar torsion box machined from tapering panels. The leading-edge honeycomb flaps and single-slotted trailing-edge flaps are electrically actuated, as are the all-metal two-section slotted spoilers.

Landing gear

The hydraulically-retractable tricycle undercarriage has pneumatic back-up for emergency actuation. The main units, each of which has side-by-side mainwheels, retract forward, while the single nosewheel retracts backward. The aircraft has hydraulic brakes and Hydro-Aire anti-skid units. The nosewheel has a 14-ply 18 x 5.5 Type VII tyre, inflated to 215 lb/sq in (14.82 bar), while the mainwheels have 18-ply 25 x 6.75 Type VII tyres inflated to 300 lb/sq in (20.69 bar). The undercarriage is extremely rugged, and allows limited rough field operations.

Ventral fins

A pair of canted ventral fins enhance directional stability by increasing keel area and by reducing the tendency to Dutch roll at transonic speeds.

Weapon options

Weapon options include JLAU-3A rocket pods (with 19 70-mm rockets), RL-7 (seven 70-mm rockets) or RL-4 (four 125-mm rockets) rocket pods, or a range of freefall bombs of up to 750 lb weight. The aircraft is also cleared to carry J/ALQ-6 ECM pods.

Internal armament

Whereas the SEPECAT Jaguar has a pair of 30-mm ADEN or DEFA cannon, the similarly roled and similarly configured F-1 follows US practice and uses a single 20-mm JM61 six-barrelled cannon. This is a licence-built version of the General Electric M61A1 Vulcan cannon employing the principle invented by Dr Gatling. The six rifled barrels are rigidly clamped together and rotate with the entire breech rotor. Each barrel has cam followers which successively chamber, fire and extract the rounds as the barrels rotate. The weapon has a high rate of fire (up to 6,000 rpm) and an impressive muzzle velocity (1036 m/s), and the rotating barrels give long life by reducing heat damage and erosion. In the F-1 the cannon uses a linkless ammunition feed system from the 750-round magazine, since belt-fed systems have proved unreliable when used with this type of weapon. The linkless feed system consists of a conveyor belt in a flexible housing, with rounds stored in radial partitions within the drum. A central rotor, fashioned like an Archimedes screw, drives the rounds from the drum and along the conveyor belt.

Engine intakes

The Mitsubishi F-1's twin Adour engines are fed by separate square-section intakes mounted quite high on the forward fuselage. These are fitted with fixed intake ramp/splitter plates, which prevent sluggish boundary layer airflow from entering the intake. These splitter plates are similar to those originally fitted to early prototype Jaguars. The intakes have spring-loaded suck-in auxiliary intake doors aft of the lip, which is usually unpainted, and which is electrically de-iced. The simple fixed geometry intake saves weight and complexity, but limits speed to well under Mach 2. Prominent triangular warnings are painted on the intake sides.

Arrester hook

The F-1 is fitted with a spring-loaded arrester hook on the keel between the two engine nozzles.

RHAWS

A fin-tip fairing houses the forward and rear hemisphere antennas for the Tokyo Keiki J/APR-3 RHAWS (Radar Homing And Warning System). The transmitter/receiver computer is located in the avionics bay which replaces what had been the rear cockpit on the two-seater, along with the J/ASN-1 Inertial Navigation System (an indigenous version of the Ferranti 6TNJ-F INS), the J/A24G-3 Air Data Computer and the J/ASQ-1 Weapons Release Computer, all of which are unique to the F-1 in JASDF service.

Fuel dump

The fuel dump pipe projects from the trailing edge of the tailfin, immediately above the top of the rudder. The same fairing also mounts the rearward-facing white navigation light.

Powerplant

Like the T-2, the F-1 is powered by a pair of Ishikawajima-Harima Heavy Industries TF40-IHI-801A augmented turbofans, rated at 7,305 lb st (32.49 kN) with afterburning, and at 5,115 lb st (22.75 kN) in maximum dry thrust. This engine is a licence-built version of the Rolls-Royce Turboméca Adour Mk 801 and delivers the same thrust as the Adours fitted to French Jaguars, but less than those fitted to British and export Jaguars (405 lb less in dry thrust, and 1,095 lb in maximum reheat). The Jaguar's maximum take-off weight is 4,409 lb (2000 kg) greater than that of the F-1. This gives the F-1 a performance slightly superior to that of the French Jaguar, and slightly inferior to that of the Jaguar International. The lack of part-throttle reheat is a major disadvantage by comparison with the British and export Jaguars.

Brake parachute

The tailcone contains a single 18-ft diameter ring-slot-type braking parachute. This augments the powerful Sumitomi Precision Metal Company anti-skid brakes, and the airfield arrester hook stowed on the keel between the engine nozzles.

Specification
Mitsubishi F-1

Powerplant: two Ishikawajima-Harima TF40-IHI-801 (licence-built Rolls-Royce/Turboméca Adour Mk 801A) augmented turbofans, each with a max sea level ISA thrust of 5,115 lb st (22.75 kN), or 7,305 lb st (32.49 kN) with afterburning
Wingspan: 7.88 m (25 ft 10 in)
Wing chord (root): 4.172 m (13 ft 8¼ in)
Wing chord (tip): 1.133 m (3 ft 8½ in)
Wing area: 21.17m² (227.9 sq ft)
Aspect ratio: 3
Taper ratio: 3.7
Leading-edge flap: operating angles 15° and 30°, area 1.54 m²
Trailing-edge flap: operating angles 25° and 40°, area 2.61 m²
Tailplane span: 4.33 m (14 ft 2½ in)
Tailplane area: 6.70 m² (72.12 sq ft)
Overall length: 17.86 m (58 ft 7 in)
Fuselage length: 17.31 m (56 ft 9½ in)
Height: 4.45 m (14 ft 7 in)
Fin area: 5.00 m² (53.82 sq ft)
Wheel track: 2.82 m (9 ft 3 in)
Wheelbase: 5.72 m (18 ft 9 in)
Empty weight: 6358 kg (14,017 lb)
Weight equipped: clean, full fuel with only 200 rounds 9690 kg (21,362 lb); air-to-air training with two AAMs 9880 kg (21,781 lb); air-to-ground with two 750 lb bombs, three external tanks and 500 rounds 13,000 kg (28,660 lb)
Maximum take-off weight: 13700 kg (30,203 lb)
Maximum internal fuel: 3823 litres (1,010 US gal; 841 Imp gal)
Maximum external fuel: 2463 litres (651 US gal; 540 Imp gal)
Maximum external load: 2721 kg (12,500 lb)
Maximum level speed: Mach 1.6
Stalling speed: 222 km/h (138 mph; 117 kt) with gear and flaps down
Maximum initial climb: 177 m/sec (35,000 ft/min)
Service ceiling: 15240 m (50,000 ft)
Take-off run: 610 m (2,000 ft) SL/ISA clean; 914 m (3,000 ft) with two 750 lb bombs and one external fuel tank
Landing roll: 850 m (2,800 ft) with two 750-lb bombs and one external fuel tank; 548 m (1,800 ft) with AAMs only
Combat radius: Intercept role, two AAMs only 278 km (172 miles); air-to-surface role, with eight 500-lb bombs and two external tanks 351 km (218 miles); anti-ship with two ASM-1s and one centreline tank 556 km (345 miles)
Maximum ferry range (max external fuel): 2870 km (1,783 miles)

Colour scheme

Most F-1s wear a three-tone camouflage on their upper surfaces, with light grey undersides. The camouflage is applied in dark green and olive drab over a light brown/tan colour, and is similar to the scheme originally worn by JASDF RF-4 Phantoms. All demarcation lines are soft. A variety of experimental camouflage colour schemes has been tried, but none has been widely adopted. National insignia are applied in six positions, above and below each wing, and on the sides of the forward fuselage. Unit markings, when worn, are carried on the sides of the tailfin.

Mitsubishi F-1
3 Hikotai (Third Squadron)
3 Kokudan (Third Air Wing)
Hokubu Koku Homentai
(Northern Air Defence Command)
Misawa Air Base,
Honshu, Japan

The Mitsubishi F-1 retains a close similarity to the T-2 trainer – from which it was derived – in appearance, structure and equipment. The aircraft also bears more than a passing resemblance to the Anglo-French SEPECAT Jaguar, which Japan had wanted to build under licence. This aircraft wears the distinctive Kabuto (Samurai helmet) markings of the 3rd Hikotai, with twin horizontal tail bands and nose art applied for participation in a 1994 ACM/gunnery meet. Named 'Katsu' on one side of the nose, the aircraft also carried the legend 'Fly with Runner' on the opposite side. The same motto was carried by four more of the team aircraft during the competition. Such nose art is almost inevitably applied in English, sometimes with the words painted on reversed! These competition markings are inevitably short-lived, although frequent competitions and fighter meets mean that each aircraft may be so decorated several times during its flying career. The 3rd Hikotai is normally based at Misawa in the far north of the largest Japanese island of Honshu, a base it shares with a second Mitsubishi F-1 squadron, and with units equipped with the F-15J Eagle, the E-2C Hawkeye, and with the USAF's 35th Fighter Wing which flies F-16Cs from the same airfield.

Radar
The Mitsubishi F-1 is fitted with a Mitsubishi-Denki J/AWG-12 fire control system, which is basically similar to the J/AWG-11 fitted to the armed two-seat T-2K variant. The J/AWG-12 is a Japanese-built version of the AN/AWG-12 fire control system fitted to RAF F-4M Phantom FGR.Mk 2s, while the J/AWG-11 was based on the AN/AWG-11 fitted to the Royal Navy's F-4K Phantom FG.Mk 1. The AN/AWG-12 was built around the Westinghouse AN/APG-61 radar, and it is assumed that this radar also forms the core of the Japanese fire control system. The radar is thus extremely primitive and old fashioned, and is similar to the AN/APG-59 radar of the basic AWG-10 fire control system fitted to US Navy F-4Js. This could claim to be the world's first fire control system to use transistorised circuitry, and offered a limited look-down/shoot-down capability, albeit at fairly short ranges. It is believed that the J/AWG-12 has the upgrades built into the British AWG-11 and AWG-12 and applied to most AN/AWG-10s during the mid-1970s. These substituted a solid state transmitter, using a klystron power amplifier, and also added a digital computer. Three LRUs were modified, six were replaced by newly designed equipment and seven were deleted altogether. The radar presumably has expanded capabilities in the sea-search mode, although it falls far short of dedicated maritime attack radars like the Agave (fitted to Indian maritime Jaguars and to various Mirage sub-variants). The J/AWG-12 fire control system and associated radar were reportedly selected in preference to an indigenous radar which would have been produced by Toshiba.

Radome
The flat plate antenna of the J/AWG-12 is covered by a sharply pointed glossy black radome, tipped by a conventional pitot-static probe. The small base diameter of the radome would seem to indicate that the antenna itself is relatively small, which would in turn suggest a rather limited range. The aircraft does not have an illuminator for missile guidance, and is not believed to have any automatic terrain-following/terrain-avoidance capability.

Anti-ship missile
The Type 80 ASM-1 was Japan's first indigenous air-to-surface missile, and was developed in a programme which began in 1973. The missile, which employs inertial guidance and active radar terminal homing, has a radio altimeter to control its sea-skimming profile, and is powered by a solid fuel rocket motor. It has a range of about 50 km. The missile has four fixed delta wings and four-tail mounted triangular control fins, and is fitted with a 150-kg semi-armour-piercing warhead. This makes the missile broadly equivalent to the Norwegian AGM-119 Penguin and the French AM39 Exocet, and similar to the smaller, shorter-range German Kormoran, although the latter missile has a significantly larger warhead. The British Sea Eagle and Taiwanese Hsiung Feng 2 use turbofan engines, and thus have rather longer reach. Prototype ASM-1 missiles were flown in 1977, and were evaluated by the JASDF during 1981 and 1982. It entered service in 1983 and has been cleared for carriage by the F-1, the P-3C Orion, and the F-4EJ Phantom. Production continues at the rate of about 30 missiles per annum. The Type 88 ASM-2 is under development as a replacement for the ASM-1, and this new missile may enter service with the F-1 fleet before the aircraft type is replaced by the FS-X. There have been reports that one squadron of F-1s will be replaced by a squadron of anti-ship-roled F-4EJ Kai Phantoms.

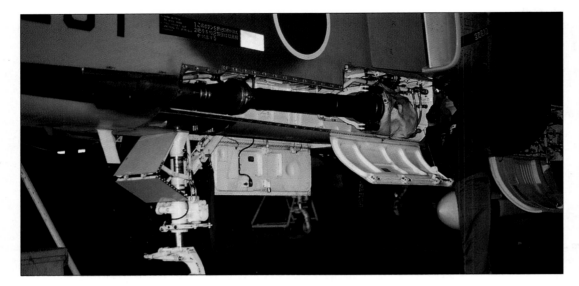

Left: The General Electric M61A1 is the JASDF's standard aircraft cannon, and is fitted to the F-15, the F-4EJ and the F-1. The same weapon was also fitted to the F-104J. The M61A1 is an electrically-powered, six-barrelled Vulcan gun and offers a high rate of fire. It is very reliable, although critics maintain that the stopping power of the individual 20-mm shells is not as effective as that enjoyed by heavier calibre rounds. Most US-designed combat aircraft use the M61A1. Russian and West European practice has been to go for pairs of larger-calibre cannon, which fire slightly more slowly. The gun can be augmented by IR-homing air-to-air missiles, which can be carried on wingtip launch rails or on the outboard underwing pylons.

Inside the Mitsubishi F-1

1 Pitot tube
2 Radome
3 Radar scanner
4 Temperature probe
5 Mitsubishi Electric (J/AWG-12) dual-mode (air-to-air and air-to-ground) radar
6 Forward radar warning antenna
7 Radar altimeter
8 Sideslip indicator probe
9 Electronics cooling air ducts
10 Radio and electronics equipment
11 Liquid oxygen converter
12 Windscreen panels
13 Mitsubishi Electric (Thomson-CSF) head-up display
14 Instrument panel shroud
15 Rudder pedals
16 Angle-of-attack probe
17 Engine throttles
18 Control column
19 Starboard side console panel
20 Daiseru (Weber) ES-7J 'zero-zero' ejection seat
21 Safety harness
22 Arm restraints
23 Port side console panel
24 Cannon blast trough
25 UHF aerial
26 Nosewheel leg door
27 Torque scissors
28 Nosewheel
29 Steering motor
30 Cannon barrel fairing
31 Cockpit aft bulkhead
32 Ejection seat rails
33 Headrest
34 Cockpit canopy cover
35 Canopy jack
36 Canopy hinges
37 Electrical system junction box
38 Inertial navigation unit

39 Air conditioning ducting
40 Nosewheel bay
41 Cannon barrels
42 Computer interface unit
43 RHAWS computer
44 Electronics bay access door
45 Gun camera controller
46 Starboard engine intake
47 Ammunition drum (750 rounds capacity)
48 Ammunition feed belt
49 Boundary layer splitter plate
50 General Electric JM61A1 six-barrelled 20-mm rotary cannon
51 Port intake
52 Intake suction relief doors
53 Air conditioning plant
54 Lower identification light
55 No. 1 fuselage fuel tank, total internal capacity 841 Imp gal (3823 litres)
56 Bleed air louvres
57 Boundary layer air duct
58 Control cable runs
59 Identification light
60 TACAN aerial
61 Dorsal spine fairing
62 Fuel system piping
63 No. 2 fuselage fuel tank
64 Intake trunking
65 Port wing root leading-edge extension
66 Mainwheel (stowed position)
67 Mainwheel doors
68 Port leading-edge flap electric motor
69 Leading-edge flap motor interconnection
70 No. 3 fuselage fuel tank
71 No. 4 fuselage fuel tank

72 Wing centre-section carry through structure
73 Fuselage/front spar attachment joint
74 Wingroot extension
75 Starboard leading-edge drive motor
76 Wing fence
77 Starboard inner stores pylon
78 Pylon attaching rib
79 Starboard outer stores pylon
80 Leading-edge dog tooth
81 Missile launching rail
82 AIM-9 Sidewinder air-to-air missile
83 Starboard navigation light
84 Fixed portion of trailing edge
85 Starboard double-slotted flap
86 Roll control spoilers
87 Spoiler actuating links
88 Fuselage/rear spar attaching joint
89 Flap electric motor
90 Spoiler differential hydraulic jack
91 No. 5 fuselage fuel tank
92 Flap motor synchronising interconnection
93 No. 6 fuselage fuel tank
94 Fuselage skin plating

95 Aft identification light
96 Artificial feel system pressure sensor
97 Fuel vent valves
98 Compass transmitter
99 Tailfin construction
100 Aluminium honeycomb leading-edge panel
101 Starboard tailplane
102 Anti-collision light
103 UHF aerials
104 Rear radar warning aerials
105 Tail navigation light
106 Fuel jettison
107 Rudder construction
108 Honeycomb trailing-edge section
109 Rudder hydraulic actuator
110 Brake parachute housing
111 Tailcone
112 Port all-moving tailplane
113 Honeycomb trailing-edge panel
114 Tailplane spar construction

115 Pivot bearing
116 Tailplane hydraulic jacks
117 Tail control linkages
118 Runway arrester hook
119 Variable area afterburner exhaust nozzle
120 Fireproof bulkhead
121 Engine access doors
122 No. 7 fuselage fuel tank bay
123 Rolls-Royce/Turboméca Adour afterburning turbofan
124 Engine accessories
125 Primary heat exchanger
126 Port flap drive motor
127 Hydraulic reservoir
128 Port ventral fin
129 Port double slotted flap
130 Honeycomb trailing-edge section

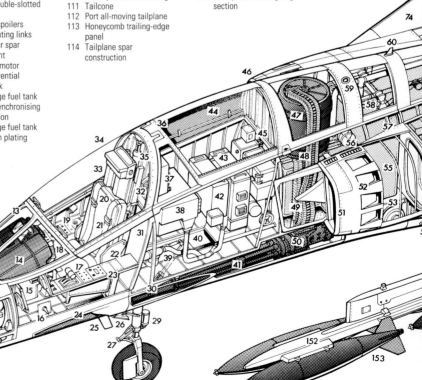

Mitsubishi F-1

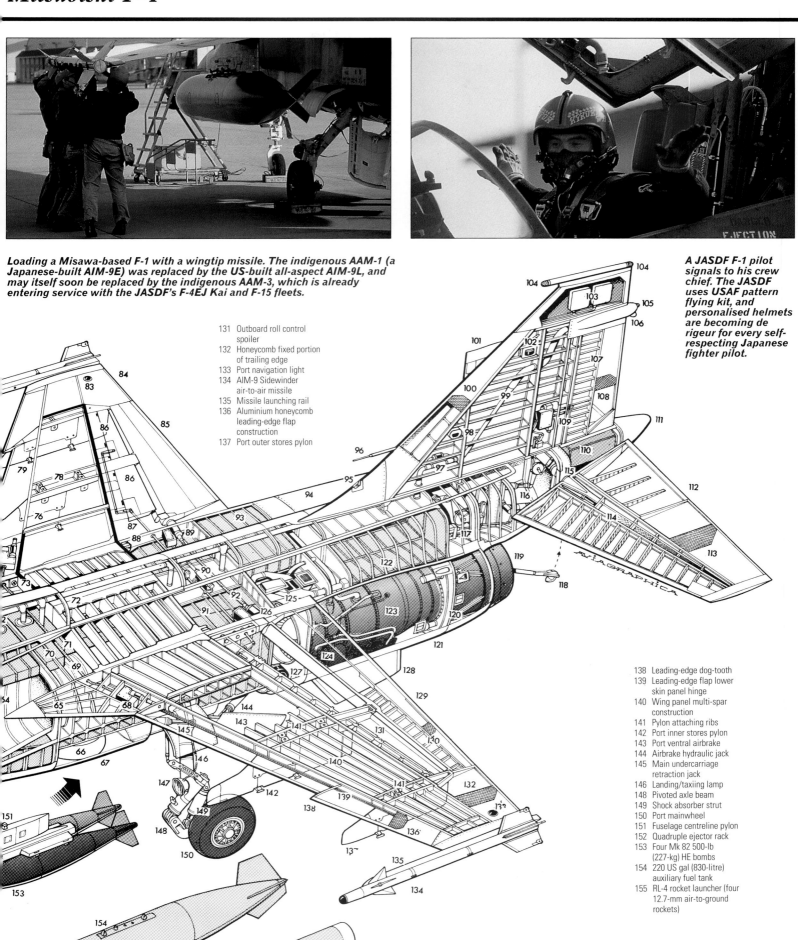

Loading a Misawa-based F-1 with a wingtip missile. The indigenous AAM-1 (a Japanese-built AIM-9E) was replaced by the US-built all-aspect AIM-9L, and may itself soon be replaced by the indigenous AAM-3, which is already entering service with the JASDF's F-4EJ Kai and F-15 fleets.

A JASDF F-1 pilot signals to his crew chief. The JASDF uses USAF pattern flying kit, and personalised helmets are becoming de rigeur for every self-respecting Japanese fighter pilot.

131 Outboard roll control spoiler
132 Honeycomb fixed portion of trailing edge
133 Port navigation light
134 AIM-9 Sidewinder air-to-air missile
135 Missile launching rail
136 Aluminium honeycomb leading-edge flap construction
137 Port outer stores pylon

138 Leading-edge dog-tooth
139 Leading-edge flap lower skin panel hinge
140 Wing panel multi-spar construction
141 Pylon attaching ribs
142 Port inner stores pylon
143 Port ventral airbrake
144 Airbrake hydraulic jack
145 Main undercarriage retraction jack
146 Landing/taxiing lamp
148 Pivoted axle beam
149 Shock absorber strut
150 Port mainwheel
151 Fuselage centreline pylon
152 Quadruple ejector rack
153 Four Mk 82 500-lb (227-kg) HE bombs
154 220 US gal (830-litre) auxiliary fuel tank
155 RL-4 rocket launcher (four 12.7-mm air-to-ground rockets)

Fuel

Unlike the SEPECAT Jaguar, the F-1 does not have a wet wing, and all fuel has to be stored in seven bag-type tanks in the fuselage or in the three external auxiliary tanks. Internal fuel capacity is 1,010 US gal (3823 litres), significantly less than that of the similar SEPECAT Jaguar, giving the aircraft a somewhat reduced radius of action. The Mitsubishi F-1's role is always said to be one of 'anti-landing craft' and the type's limited range makes this more reasonable than a true 'defence of the sea lanes' power-projection anti-shipping strike role.

AIM-9 Sidewinder

The F-1 can carry up to four AIM-9 Sidewinder IR-homing air-to-air missiles on its wingtip launch rails and on LAU7/AA launchers fitted to the outboard pair of underwing pylons. This allows the aircraft to undertake a limited secondary air defence role. At one time, consideration was reportedly given to fitting Jaguar-style overwing missile launch rails. The AIM-9E was built locally as the AAM-1, and it had been planned to produce an indigenously designed improved version as the AAM-2. This was cancelled in favour of the all-aspect AIM-9L, however, which was also built locally. The indigenous Type 90 AAM-3 IR-homing air-to-air missile entered development in 1985, and flight trials began in 1989. The missile, which has an unusual notched control fin configuration, is understood to be entering service with the F-4EJ Kai and F-15J, and may later equip the Mitsubishi F-1. It will offer even better target discrimination than the AIM-9L, while its seeker will have a wider 'look' angle. The missile will also have greater agility and acceleration, with a more lethal warhead and a more efficient rocket motor.

Spine

The Mitsubishi F-1 was a minimum-change derivative of the two-seat T-2. The same fuselage contours were used for both aircraft types, to negate the need for new aerodynamic and wind-tunnel testing, and to minimise changes to production tooling. The redundant rear cockpit was used as an avionics bay, and the rear canopy was replaced by a metal access hatch. The end result is that the F-1 has a distinctly hump-backed appearance, with higher drag than would strictly be necessary.

Ejection seat

The F-1 uses a Daiseru-built version of the American Weber ES-7J zero-zero ejection seat. This is fitted with locally designed canopy penetrators, allowing the pilot to eject through the canopy in extremis. This is a primitive but cost-effective solution to the problem of minimising the delay between seat actuation and escape, without having to use explosive MDC (Miniature Detonating Cord) embedded in the canopy.

Mitsubishi F-1 operators

The Mitsubishi F-1 has replaced the F-86F Sabre in the tactical fighter role with three JASDF squadrons, two of these serving with the 3rd Kokudan at Misawa as part of the Northern Air Defence Force, and one with the 8th Kokudan at Tsuiki as part of the Western Air Defence Force. Following conversion, on the production line, of the sixth and seventh T-2s as prototypes (designated FS-T-2 Kai), production of the Mitsubishi F-1 reached 77 aircraft. Blocks of these aircraft were interspersed on a common production line with batches of two-seat T-2 trainers. Details of T-2/F-1 production are given below, with F-1 serials and construction numbers in bold type.

C/N	SERIAL NUMBERS	TYPE
01-04	19-5101, 20-5102 to 5104	T-2P
05	59-5105	T-2Z
06-07	**59-5106 and 5107**	**FS-T-2 Kai**
08-16	59-5108 to 59-5116	T-2Z
17-24	69-5117 to 69-5124	T-2Z
25-35	69-5125 to 69-5135	T-2K
36-46	79-5136 to 79-5146	T-2K
47-55	**70-8201 to 70-8209**	**F-1**
56-64	**80-8210 to 80-8218**	**F-1**
65-74	89-5147 to 85-5156	T-2Z
75-79	**80-8219 to 80-8223**	**F-1**
80-82	**90-8224 to 90-8226**	**F-1**
83-89	99-5157 to 99-5163	T-2K
90-97	**90-8227 to 90-8234**	**F-1**
99-115	**00-8236 to 00-8252**	**F-1**
116-122	**10-8253 to 10-8259**	**F-1**
123-133	19-5164 to 19-5174	T-2K
134-136	29-5175 to 29-5177	T-2K
unknown	29-5178	T-2K
unknown	20-8260 to 20-8266	F-1
unknown	39-5179 to 39-5182	T-2K
unknown	30-8267 to 30-8269	F-1
unknown	49-5184 to 49-5190	T-2K
unknown	**50-8270 to 50-8271**	**F-1**
unknown	**60-8272 to 60-8274**	**F-1**
unknown	79-5193 to 79-5194	T-2K
unknown	**70-8275 to 70-8277**	**F-1**
unknown	89-5195 to 89-5196	T-2K

Several Mitsubishi F-1s have worn sharkmouths during the type's service career, but such markings have been short-lived. Until 1994, there was a superstition that sharkmouthed aircraft could not win gunnery and ACM competitions. A sharkmouthed F-1 disproved the folklore! This 3rd Hikotai aircraft was seen at Komatsu during November 1984, and is fitted with a centreline practice bomb carrier.

Two Mitsubishi F-1s (10-8257 and 10-8258) are believed to have been relegated to ground instructional status with the 1st Technical School (part of the Jyutsuka Kyoiku Honbu, or Technical Training Command) at Hamamatsu, but may be held as an attrition reserve. At least two more have been lost in accidents (00-8237 on 20 August 1980 and 80-8218 on 10 April 1987). Most of the remaining 73 aircraft are divided between the three front-line squadrons, each of which also have a handful of T-2s on charge for conversion, continuation and instrument training, and for use as hacks. It is uncertain as to exactly how many F-1s are currently undergoing a Service Life Extension Programme (SLEP), although the number will be sufficient to equip two of the three squadrons for several more years. It has not been decided which of the three current F-1 squadrons will be disbanded when the force contracts.

Japan's anti-militarist stance has made it difficult to export military aircraft, and no foreign customers have been found for the F-1. All operators of the type are thus JASDF units, and these are detailed below.

Hokubu Koku Homentai (Northern Air Defence Command)

3 Kokudan (Third Air Wing)

3 Hikotai (Third Squadron), Misawa Air Base

The Third Hikotai served for more than 20 years as a tactical fighter squadron equipped with the F-86F Sabre. It formed on 1 October 1956 at Hamamatsu as part of the 2nd Kokudan, officially moving to Chitose on 2 September 1957, three and a half months after its aircraft and crews actually relocated. The unit transferred to the 4th Kokudan at Matsushima on 5 March 1963, and to the 81st Kokutai at Hachinoe on 1 February 1964. The unit moved to Misawa on 1 March 1971 and started conversion to the Mitsubishi F-1 in September 1977. The squadron transferred to the 3rd Kokudan on 1 March 1978, with re-equipment complete. Until 1983, aircraft assigned to the 3rd Kokudan wore the wing insignia, which consisted of a stylised red number 3, with an outline of the Aomori Prefecture, in which Misawa was located. Individual squadron markings were adopted from 1983. The 3rd Hikotai adopted a Kabuto (Samurai's helmet) badge in August 1983.

Mitsubishi F-1:
(Those listed below with names and artwork participated in the 1994 Senkyo at Komatsu)
80-8223 (Warrior insignia)
90-8230 Fly with Runner
90-8231 (white Disney 'Alice in Wonderland' rabbit insignia)
90-8233 Ace/Break the Target with Bomb and Gun
00-8235 Cobra Samurai/Fly with Runner
00-8242 Better watch out
00-8246 River/Bomb Girl
00-8261 Jun/Fly with Runner
30-8269 Formidable Bird/Hit the Bullseye
50-8270 Kid
60-8272 Katsu/Fly with Runner
70-8201, 70-8203, 70-8204, 70-8209, 80-8213, 80-8215, 80-8218, 80-8223, 90-8225, 90-8228, 90-8231, 90-8233, 90-8234, 00-8235, 00-8245, 00-8250, 10-8253, 10-8257, 10-8258, 10-8259, 20-8261, 20-8262, 20-8264, 20-8265, 30-8267, 30-8269, 50-8270, 60-8272, 60-8273, 60-8274,
Mitsubishi T-2:
59-5111 (grey colour scheme), 59-5112 (grey colour scheme), 49-5183 (grey colour scheme), 49-5186

Left: The traditional Kabuto (Samurai warrior's helmet) insignia of the 3rd Hikotai was adopted in August 1983. It is usually applied to the tailfin in glorious technicolour, though some toned-down aircraft have worn outline versions of the badge, or with the badge lightly overpainted to reduce its conspicuity.

Above: The stylised 3 used by the 3rd Kokudan as a unit badge lost some of its point when applied 'mirror fashion' on the starboard side of the tailfin.

Below: One of the 3rd Hikotai's toned-down F-1s sports a coat of sea grey paint that all but covers the original camouflage, and has small, toned-down Hinomaru.

8 Hikotai (Eighth Squadron)

The Eighth Hikotai began conversion to the Mitsubishi F-1 on 30 June 1979, becoming the second of the 3rd Kokudan squadrons to re-equip with this potent maritime strike aircraft. The squadron had originally formed on 29 October 1960 at Matsushima as part of the 4th Kokudan, and equipped with the F-86F. The unit transferred to the Rinji Komatsu Hakentai (Komatsu Provisional Air Group) on 1 May 1961. This became the 6th Kokudan on 15 July 1961. The squadron transferred to Iwakuni on 25 November 1964 and formed the nucleus of the 82nd Kokutai, which was formed on 1 December 1964, disbanding on 1 December 1967. The 8th Hikotai then transferred to the 3rd Kokudan at Komaki, and moved to Misawa on 31 March 1978. For the first few years of its existence, the 8th Hikotai's Mitsubishi F-1s wore the emblem of the 3rd Kokudan on their fins. This was applied in yellow from 1980, in order to differentiate between aircraft assigned to the 3rd and 8th Hikotai. When individual squadron markings were adopted in 1983 the 8th Hikotai (nicknamed the 'Panthers') adopted a fearsome black panther badge designed by a squadron member, T/Sgt Yasuda.

Mitsubishi F-1

(The aircraft listed below with markings descriptions participated in the 1994 Senkyo at Chitose. The aircraft wore a kanji character on the port side of the rudder, with a huge black and white painting on the starboard side. The figures represented characters from the Edo period (1603-

1863) novels of Bakin Takizawa, collectively known as the Nanso Satomi Hakken Den – the Legend of the Eight Dogs of Nanso).

70-8201 (Jin – charity)
70-8202 (Chu – loyalty)
70-8203 (Kenzan – come and see)
80-8220 (Gi – justice)
90-8234 (Tei – obedience)
00-8240 (Ko – filial piety)
00-8241 (Shin – trust)
10-8258 (Rei – respect)
20-8262 (Chi – wisdom, with sharkmouth)
60-8273 (sharkmouth)
70-8204, 70-8205, 80-8210, 80-2212, 80-8220, 80-8221, 80-8222, 90-8224, 90-8225, 90-8226, 90-8230, 90-8232, 90-8234, 00-8236, 00-8238, 00-8240, 00-8242, 00-8244, 00-8246, 00-8247, 254, 20-8261, 20-8262, 60-8274

Mitsubishi T-2

69-5120 (grey colour scheme), 89-5148 (grey colour scheme), 59-5191

Above: A Mitsubishi F-1 of the 8th Hikotai is seen on approach, laden with underwing fuel tanks. The F-1 is hampered by its poor range and warload characteristics.

Right: An interesting non-standard adaptation of the 8th Hikotai black panther badge, with the animal clawing a red star. The badge was designed by a squadron member and replaced a stylised yellow 3. The panther faces forward on both sides of the tailfin.

Right: The tailfin of an 8th Hikotai F-1 shows the normal serial and badge presentation and position, and the normal camouflage colours.

Below: This aircraft has had its squadron badge, nose code and serial virtually obscured, and has reduced-size national insignia, too.

Seibu Koku Homentai (Western Air Defence Command)

8 Kokudan (Eighth Air Wing)

6 Hikotai (Sixth Squadron), Tsuiki Air Base

The Sixth Hikotai was formed at Chitose on 1 August 1959, with F-86Fs, and partnered the Third Hikotai within the 2nd Kokudan until 1 April 1960 when it transferred to 5th Kokudan, which had formed at Matsushima on 1 December 1959. The squadron had moved to Nyutabaru on 1 October 1959, where the 5th Kokudan HQ was relocated from 4 July 1960. The squadron transferred to the Rinji Tsuiki Hakentai (Tsuiki Provisional Air Group) on 26 October 1964, this unit becoming the 8th Kokudan on 28 December 1964. The squadron began conversion to the Mitsubishi F-1 on 11 March 1980, losing its last pair of F-86Fs in November. The squadron was the last to re-equip with the Mitsubishi F-1, after the two 3rd Kokudan units, which gave the Western Air Defence Force its first true anti-ship capability. The Mitsubishi F-1s of the 6th Hikotai have always worn a squadron badge, and have never been decorated with the insignia of the parent

8th Kokudan. The squadron insignia consists of three traditional Samurai weapons superimposed on a yellow disc.

Mitsubishi F-1
70-8206, 70-8209, 80-8211, 80-8215, 80-2217, 80-2219, 80-2222, 90-8227, 90-8229, 90-8231, 90-8233, 00-8236, 00-8239, 00-8240, 00-8241, 00-8248, 10-8251, 10-8256, 20-8263, 30-8268, 50-8271, 70-8276, 70-8277

Mitsubishi T-2:
59-5192

Below: This 6th Hikotai F-1 was specially decorated for the 50th anniversary of Tsuiki, with the various based types (Imperial Japanese, USAF occupation and JASDF) represented on the belly tank.

Above: Two 6th Hikotai tailfins show the variation in colour schemes common during the late 1980s. The far aircraft wears standard markings, while the nearer machine, in a temporary maritime camouflage scheme, has an extra 'shooting star' badge.

Research and Development Command

Koku Jikkenden (Air Development and Test Wing), Gifu

The two Mitsubishi FS-T-2 Kai prototypes continue to serve with the ADTW at Gifu, alongside several of the early T-2 prototypes, and have occasionally been joined by F-1s borrowed from the front-line squadrons. The prototypes have not been brought up to full F-1 standards, and remain painted in their original grey and orange colour scheme, but are nevertheless extremely useful development and test mules. The unit's permanent aircraft wear a badge consisting of a blue chevron, representing a supersonic shock wave, with the letters ADTW. These letters replaced APW (for Air Proving Wing) during the 1980s, when the unit was redesignated.

Mitsubishi FS-T-2 Kai:
59-5106, 59-5107

Right: The first of the two Mitsubishi FS-T2 Kai prototypes wears Air Proving Wing markings at Gifu. The metal rear cockpit cover, fixed inside a normal canopy, is noteworthy.

In the secretive world of special operations, misconceptions can easily arise. Mistrust of the unconventional warrior remains endemic and has often camouflaged AFSOC's professionalism and combat record. After years as poor relations, parented by formations with no real knowledge of their mission, Special Operations Forces have at last been concentrated in an independent command, ready to fight on their terms, any time, any place.

On the night of 24 April 1980, in the desolate sands of Iran's Dasht-e-Kavir, more was lost than the lives of eight brave men. As Americans watched the televised images of Iranians desecrating the bodies of the servicemen killed at 'Desert One' – during the abortive hostage rescue attempt – the country's pride, confidence and prestige plummeted. An unprecedented sense of ineptitude swept the nation, along with the existing frustration over the fate of the 53 American hostages held at the US Embassy after Ayotollah Khomeni's Islamic revolution of 1979. Cynics dubbed Colonel Beckwith's brave rescuers the 'Delta Farce' and the whole concept of special operations seemed to be discredited. Indeed, in no small degree, the fiasco swept the Carter administration from office.

The phoenix that is now known as the United States Special Operations Command (SOCOM) rose out of the ashes of this defeat and humiliation. It has since emerged as one of the best trained, most highly specialised and respected commands in the US military, and is steadily overcoming a residue of ill-informed criticism and a range of jealousies. With future enemy threats most likely to come from petty despots located halfway around the world, the uniquely trained and highly motivated special operations forces of USSOCOM have become increasingly vital to US interests, not least because they are the only element of the US military that is equipped, trained and ready to respond immediately, any time, any place.

Following the fiasco at 'Desert One', a panel was appointed to study the mission, report on why it failed and recommend ways on how to ensure success in the future. Chaired by former Chief of Naval Operations Admiral James L. Holloway, the panel was made up of retired and active-duty generals and flag officers. The 'Holloway Report' made two recommendations. The first was to form

This 1st SOS Mod 90 MC-130E has none of the obvious external features, such as Fulton gear or a modified nose, which characterise earlier Combat Talon Is. Only more subtle modifications – its FLIR and RWR antennas, coupled with a ghostly colour scheme – give any clues to its mission.

AFSOC

United States Air Force
Special Operations Command

US Air Force Special Operations Command

Early models of the MC-130E, such as this MC-130E-C (Clamp), were typified by their forked Fulton STAR recovery gear and the reprofiled nose, housing an APQ-122 terrain following radar. An aircraft rigged for a Fulton recovery mission was fitted with an array of wires to guide the 'snatched' subject away from the propellers and tail if the 'prongs' failed to snag the nylon cable connecting him to the recovery balloon. The first 'live' Skyhook pick-up was carried out on 24 August 1966.

The MC-130Es used by the 8th SOS still carry Fulton equipment, though it is no longer a prime mission for the unit. In the years since the Combat Talon I entered service it has undergone many modernisation programmes and aircraft are currently undergoing the Mod 90 upgrade. This entails a substantial refit of the avionics and electrical systems along with changes to the main cargo cabin.

a Counter-Terrorist Joint Task Force (CTJTF) as a field agency of the Joint Chiefs of Staff (JCS), with its own permanently-assigned staff personnel and certain assigned forces. The second recommendation suggested that the JCS should consider establishing a Special Operations Advisory Panel, comprised of high-ranking active-duty and retired officers with special operations backgrounds, or who had served at the Commander-in-Chief (CINC) or JCS levels and who had maintained an ongoing interest in special operations matters.

While the Defense Department accepted and implemented both recommendations in autumn 1980, practitioners of unconventional warfare were still far from having a fully autonomous special operations forces command, a goal which still lay several years in the future. Traditional military commanders have long been distrustful of special operations warriors, whom they viewed as 'cowboys' or 'snake eaters', and more than a few generals and admirals felt that special operators were seeking and getting too much independence. Few accepted the idea that autonomous special operations forces could ever constitute a viable entity.

Laying the ground for SOCOM

Against this hostility and lack of understanding, and in spite of wartime experience in Vietnam, by the mid-1980s US Army Special Forces – the Green Berets of lore – had dwindled to three active-duty groups, from a wartime high of seven. Similarly, the Air Force special operations AC-130 gunships were scheduled for transfer to the Reserves and the ageing fleet of MC-130 Combat Talons was in decline, due to a lack of significant modifications. Although the Goldwater-Nichols Act of 1984 rewarded officers for fostering inter-service co-operation – what is known as 'jointness' – and penalised them for parochial loyalty to their own branches, the legislation did nothing to encourage duty in the special operations field. Soldiers, sailors and airmen believed – often accurately – that accepting a special opera-

tions posting amounted to throwing a career away. Morale among special operations forces had hit an all-time low.

The activation of the special unit proposed by the Holloway report was at least a start. It was a first step and signalled the beginning of a shift in attitudes within the military toward the importance of special operations forces.

By the middle of 1984, Congressional interest in SOF had increased. The 23 October 1983 bombing of the Marine barracks in Beirut and the Grenada operation in November 1983 both focused Congressional attention on terrorism and low-intensity conflict. Many decision-makers started to question what roles SOF could or should play. Eventually, after much prodding, education and hard work

by a number of visionaries in both the military and civilian sectors, the US Congress made the bold and unprecedented move of passing the Cohen-Nunn Act. This created the United States Special Operations Command (USSOCOM), a unified command headed by a four-star Army, Marine Corps or Air Force general or Navy admiral. The commander is appointed politically and there is not a formal, structured rotation between the services. USSOCOM came into existence on 16 April 1987. The current Commander-In-Chief is General Wayne A. Downing, a seasoned US Army veteran of unconventional warfare.

USSOCOM brings a new dawn

With the creation of USSOCOM, a renewed pride and *esprit de corps* emerged at the operational level. No longer was a career in special forces automatically considered to be a dead end by senior military officers. SOF began to attract its share of the best and the brightest. Training took on an intensified urgency as funding increased and as interest at the highest levels of government focused on special operations. The key question being asked was how the Air Force, Army and Navy special operations people might best work effectively together.

USSOCOM's mission is to prepare special operations forces for worldwide special and psychological operations in peace and war. It also provides combat-ready special operations forces to the other unified commands as needed. The Command is responsible for all joint SOF training, and for developing the doctrine, tactics, techniques and procedures that ensure close co-operation and co-ordination between the separate services. The Command's three unique missions include professionally developing special forces personnel; developing and acquiring materiel, equipment and services uniquely suited for SOF activities; and, finally, consolidating

and submitting programme and budget proposals for Major Force Program 11. Congressional funding for the Department of Defense is given in 11 categories, known as Major Force Programs. SOF funds come under Major Force Program 11.

The Command has many varied missions. The Unconventional Warfare (UW) mission includes a broad spectrum of military and paramilitary operations in enemy-held, enemy-controlled or politically sensitive territories. It may include long-duration, indirect activities including the support of guerrilla warfare and other offensive, low-visibility or clandestine operations. These activities are mostly conducted by indigenous forces – organised, trained, equipped, supported and directed in varying degrees by US personnel.

Direct Action (DA) missions are designed to seize, damage or destroy targets, or to capture and recover personnel or material in support of strategic or operational objectives. DA missions are typically of short duration and involve small-scale offensive actions. They can include raiding, ambushing and direct assaults; placing mines and other munitions; conducting stand-off attacks by firing from air, ground or maritime platforms; designating or illuminating targets for precision-guided munitions; supporting cover and deception operations; or conducting independent sabotage operations inside enemy-held territory.

Special Reconnaissance (SR) missions are conducted to provide information concerning enemy capabilities, intentions and activities in support of conventional forces. The reconnaissance missions also provide target acquisition, area assessment, and post-strike data.

Foreign Internal Defense (FID) missions are designed to assist other (friendly) governments in any action taken to free and protect their societies from subversion, lawlessness and insurgency. These missions can often include the training of military and para-military forces.

Despite years of systems improvements, which have brought them to a similar level of capability as the later Combat Talon II, the fact remains that the USAF's MC-130Es are elderly aircraft from a structural point of view. The MC-130H Combat Talon II was once intended to replace these older Talon Is, but such is the demand for AFSOC's services that the newer aircraft will now only augment the work done by the ageing MC-130Es. Today, 14 MC-130Es still serve with the 8th SOS, at Hurlburt Field, and with the 1st SOS, at Kadena AFB. Hurlburt operates all the Fulton-equipped aircraft.

A Mod 90 Combat
Talon I cruises off the
coast at Dustin, Florida,
with Fort Walton visible
in the background. The
aircraft is an MC-130E-Y,
one of those usually
assigned to the 1st SOS
at Kadena. Infiltrating
special forces at night
and low-level, over
water, is a prime MC-
130E mission. Personnel
can be delivered by
static-line parachute
drop or can be simply
rolled straight out of the
back of the aircraft in
their Zodiac inflatable
boats. Nine of AFSOC's
MC-130Es are fitted
with Fulton gear, the
other five being
MC-130E-Ys or
MC-130E-Ss which
never had Fulton. Most
have now been fitted
with a refuelling
receptacle behind the
flight deck. Ultimately,
all will be modified to
allow them to carry
underwing refuelling
pods, similar to those
carried by the HC-130
Combat Shadows.

Counter-terrorism (CT) missions are designed to pre-empt or quickly resolve terrorist incidents. Psychological Operations (PSYOPS) missions are conducted to induce or reinforce foreign attitudes and behaviours that are favourable to US objectives. They are targeted at foreign governments, formal organisations or informal groups, as well as at individuals.

Civil Affairs (CA) operations are conducted as stand-alone or joint operations to establish, maintain, influence or exploit relations among military forces, civil authorities and civilian populations to facilitate military operations. In some cases, especially where the necessary infrastructure is non-existent, military forces may assume the functions that are normally the responsibility of local governments.

USSOCOM's components include the Joint Special Operations Command (JSOC), Army Special Operations Command (USASOC), the Naval Special Warfare Command (NAVSOC), and the Air Force Special Operations Command (AFSOC). Additionally assigned to USSOCOM are the John F. Kennedy Special Warfare Center and School, the US Air Force Special Operations School, and the Naval Special Warfare Center.

MAC 'special ops'

The April 1987 Cohen-Nunn law and the establishment of SOCOM were welcomed by those who had staked a career on special operations, but were criticised by traditional officers who saw the special operations community becoming, in effect, a sixth service branch of the US armed forces. In the US Air Force, the special operations people did not immediately gain their own command and remained subservient. 23rd Air Force headquarters moved to Hurlburt Field but remained a numbered air force under Military Airlift Command (MAC).

In a controversial move which accompanied the creation of SOCOM in 1987, the Air Force placed the aircraft and people of its combat search and rescue forces into its new 23rd Air Force, Military Airlift Command. The move was discouraging to many who had proudly dedicated their careers to the goal of saving those in peril. It was especially demoralising to some PJs, the pararescue jumpers who were always ready to kill the enemy if it would help them achieve a behind-the-lines rescue, but who felt that they were being asked to do a very different job when they became special operators. Today, PJs receive combat training which was not previously part of their specialty.

In August 1989, General Duane H. Cassidy, MAC commander (although Cassidy used the term on his correspondence, he was never a commander-in-chief, or CINC), divested all non-special operations units from 23rd Air Force. As a result, 23rd Air Force served a dual role – reporting to Military Airlift Command, but also functioning as the air component of SOCOM. Many wanted the special operators to have equal status with – not to be part of – a major command like MAC. This ambition was finally realised when the 23rd Air Force, then the US Air Force's special operations component operating under Military Airlift Command, became AFSOC.

AFSOC is born

On 22 May 1990, General Larry D. Welch, Air Force chief of staff, redesignated the 23rd Air Force as Air Force Special Operations Command (AFSOC). The new major command consisted of three wings – the 1st, 39th and 353rd Special Operations Wings – as well as the 1720th Special Tactics Group, the US Air Force Special Operations School and the Special Missions Operational Test and Evaluation Center. The Air Reserve components included the 919th Special Operations Group (Air Force Reserve) based at Duke Field, Florida and the 193rd Special Operations Group (ANG) based at Harrisburg Airport, Pennsylvania.

Air Force Special Operations Command (AFSOC) assets today comprise a fleet of extensively modified C-130 gunships, tankers and specialised insertion/recovery aircraft that operate alongside MH-53J Enhanced Pave-Low III and MH-60G Pave Hawk helicopters. AFSOC acts as the transportation arm of the USA's special operations forces and its missions include the insertion, extraction and resupply of special forces teams. It also provides fire support for both SOF and conventional forces, and can conduct aerial refuelling of SOF helicopters and PSYOP operations. It is organised into one active Special Operations Wing (SOW), two active Special Operations Groups (SOG) in Europe and the Far East, one reserve SOW, one reserve SOG operating in the PSYOPS role, and one active Special Tactics Group, which operates expeditionary airfields, conducts classified missions and participates in combat rescue missions. The Special Tactics Group includes combat control teams (CCT) and pararescue jumpers (PJs) who usually arrive on the scene prior to any other forces, survey, prepare and secure landing zones, provide air traffic control, call in air or artillery

Inside the MC-130E Combat Talon I

In 1995 nine Combat Talon Is remained in Fulton configuration and most still wore the three-tone 'lizard' camouflage scheme. This 8th SOS (the USAF's second oldest unit) aircraft is carrying an HRU (hose refuelling unit) to port. Note also the shielded open rear cargo door and the QRC-84-02A IRCM pod under the fuel tank.

Any regular Hercules pilot would find little remarkable about the basic Combat Talon I cockpit. Despite the many modifications that the aircraft has undergone, its crew still work in a 1950s-vintage cockpit. One of the few more recent additions is the radar display fitted in the pilot's console.

This is the navigation station on an MC-130E, which is situated immediately behind the co-pilot (on the starboard side). To the left is the display for the AN/APQ-122(V)8 terrain-following radar, while to the right are displays for the precision ground mapping radar and the FLIR.

Nine RWRs are visible on the extended tailcone of this MC-130E. Note how the receivers are angled and shielded from each other.

The MC-130E has eight chaff, and two flare, dispensers on each side. 30-round flare cartridges are fitted, while chaff is housed in bins inside the main cabin.

strikes and provide trauma and emergency medical care and evacuation of the wounded and injured. A Communications Squadron provides communications from the field to headquarters AFSOC and provides support to deployed Air Force special operations forces.

The US Air Force Special Operations School at Hurlburt Field, Florida, a few miles from Eglin AFB, provides special operations-related education to personnel from all branches of the Department of Defense, governmental agencies and allied nations. Subjects covered in the 13 courses range from regional affairs and cross-cultural communications to anti-terrorism awareness, revolutionary warfare and psychological operations.

These AFSOC-assigned units are backed up by another active Special Operations Wing, the 58th SOW at Kirtland AFB, New Mexico, whose task is lead-in special operations aircrew training. The 58th SOW is part of Air Education and Training Command (AETC, disrespectfully nicknamed 'Air Etc. Command') and does not report directly to AFSOC.

As a result of operational experience and a steadily increasing acquisition of specialised aircraft and equipment, AFSOC today operates an impressive array of highly-modified aircraft and expertly trained crews who proudly claim to 'own the night'. With night-vision goggles, terrain-avoiding/terrain-following radar, FLIR and GPS-aided navigation, the aircraft supporting the special operations mission are uniquely 'missionised' to carry out their tasks. Perhaps the most important aircraft in the special operations community is the C-130 Hercules, which is operated in several different configurations under a variety of acronyms, designations and codenames. The workhorse of the special operations Hercules fleet is the unarmed MC-130, now known as Combat Talon.

Combat Talon

The original Combat Talon is a highly modified subvariant of the Lockheed C-130E and was produced by converting existing C-130E transport airframes and C-130Es still on the production line. Over the years the Talon's capabilities have progressively evolved, drawing on lessons learned on numerous variants of the C-130 operated in a wide range of clandestine roles and hostile environments. Specially modified to support unconventional warfare and special operations forces worldwide, the Combat Talon I is capable of covertly penetrating a hostile environment at

With its Fulton 'whiskers' open for business, this MC-130E taxis in at Edwards AFB, while displaying some of the protective wire rigging required for each 'snatch' mission. Fulton, once regularly used by ARRS squadrons, has fallen from favour over the last 30 years and only the 8th SOS still regularly trains for this mission.

Left: A Combat Talon I makes a catch, at 150 mph (241 km/h). The 'subject' on the ground wears an (airdropped) protective suit and raises an attached line, tethered to a helium balloon.

Right: MC-130E units have a tradition of creating exotic training dummies for Fulton training, though this example is no Marilyn Monroe. 'Political correctness' has forced units to abandon some of their more outrageous, but well-loved, characters.

low altitudes and in inclement weather. In 1964, the Department of Defense established the requirements for a long-range aircraft capable of penetrating enemy defenses in support of unconventional warfare activities.

From 1964 through 1972, Project Thin Slice was initiated to support a classified Southeast Asia programme named Heavy Chain. Initially, two C-130Es were modified. Their use pioneered systems later incorporated into the Combat Talon fleet: terrain-following radar, surveillance capability and an EW suite for self-protection. In August 1966 the aircraft were renamed Rivet Yard I. They supported the special operations programme Combat Sam and featured fuel tank baffling, high-speed low-level aerial delivery system (HSLLADS), E-4 autopilot pitch channel monitor, FLIR and electronic self-protection to improve survivability in the low- to medium-threat environment. Four C-130Es were eventually modified under the programme (62-1843, 63-7785, 64-0564, 64-0565). Heavy Chain operations were terminated in October 1972 and the Rivet Yard aircraft were returned to Detachment 4 at Ontario for conversion to a more standard Combat Talon configuration, albeit without Fulton equipment, under the codename Rivet Yank. One of these aircraft was lost in February 1981 and was replaced by one of the two Rivet Swap NC-130Es (64-0571).

The Fulton system

With the C-130E transport programme in full swing, Lockheed produced 17 similarly equipped specialised C-130E-I 'Skyhook' or Rivet Clamp aircraft. These differed from other C-130Es rolling off the assembly line in being equipped with APQ-114 terrain-following radar, APS-54 threat warning system, and the unusual Fulton STARS Recovery System which used retractable nose-mounted recovery yokes to snatch passengers or packages off the ground by snagging and reeling in a 525-ft (160-m) nylon cable running from the load to be recovered to a

small helium-filled balloon which held the cable vertical. The system had first been demonstrated in August 1958, fitted to a modified Boeing PB-1 Flying Fortress, and later was employed by a modified Grumman S2F-1 Sentinel. Following trials using a C-130E fitted with a fixed, tubular recovery device at Pope AFB in 1965, it was decided to modify a small batch of aircraft for service use. Some were conversions of existing C-130E transports, others were modified on the production line. The first C-130E-I with Fulton was modified and a small squadron, the 314th Troop Carrier Wing's 50th TCS, was established at Sewart Air Force Base, Tennessee. Later in 1965, four of these specially modified C-130E-Is were deployed to Nha Trang, Republic of Vietnam. This allowed the initial 'Stray Goose' crews (special operations crews, Southeast Asia assigned) to participate in Combat Spear missions at Nha Trang Air Base, South Vietnam. In this context, the word 'Combat' simply identifies the programme as belonging to PACAF (Pacific Air Forces), just as the words 'Coronet', 'Volant' and 'Commando' refer, respectively, to Air National Guard, Air Mobility Command and AFSOC programmes.

MC-130 in Vietnam

Combat Spear missions included combat drops of personnel and equipment throughout North and South Vietnam, Laos and Cambodia. During the Vietnam War, Combat Talons were extensively involved in covert/clandestine operations in Laos and North Vietnam. They routinely flew unarmed, single-ship missions deep into North Vietnam under the cover of darkness to carry out unconventional warfare missions in support of Military Assistance Command's Special Operations Group. Little has been revealed about these flights but rumours abound, including the widespread impression that the C-130E-Is actually landed behind the lines in North Vietnam. No actual pickups with the Skyhook system were made during these missions.

The Combat Spear force was initially a detachment of the 314th Wing and its aircraft were flown periodically to Ching Chuan Kang Air Base, Taiwan, for heavy maintenance. The detachment became the 15th Air Commando Squadron on 15 March 1968 and, together with the similarly-equipped Project Duck Hook C-123 aircraft, was part of the 14th Air Commando Wing at Nha Trang. The Duck Hook element merged into the 15th Squadron in late 1968, although it retained a separate commander.

Most Combat Spear missions were routine deliveries to border-area airfields such as Khe Sanh and Nakhon Phanom, Thailand. These sorties provided transport for men and supplies to support cross-border operations. Combat Spear and Duck Hook in 1967 listed 25,000 passengers and

5,400 US tons of cargo. The use of these unconventional warfare transports in ordinary airlift work of this magnitude was deemed questionable and raised questions about the special operations function being separated from regular forces. Air Force clandestine missions into North Vietnam halted with the bombing cessation of 1 November 1968, and the Combat Spear appellation went out of use. The 15th became the 90th Special Operations Squadron on 31 October 1970 and flew the special operations C-123s and C-130E-Is until they were withdrawn from clandestine operations on 31 March 1972, 10 months before the end of the US role in the war. The Rivet Yank aircraft are also understood to have participated in Combat Spear. Four of the original C-130E-I conversions were lost during the Vietnam War. An 18th aircraft was converted to the same configuration after the Vietnam War.

Post-war updates

In 1970, an extensive modification of the airframe and a complex avionics upgrade known as Mod 70 took place. This modification, based on four years of combat experience, brought the Combat Talon fleet up to date in state-of-the-art electronics. As with so much of the equipment and so many of AFSOC's aircraft, the unconventional warfare transport achieved high drama in the November 1970 attempt to rescue American prisoners of war at Son Tay, North Vietnam – a mission for which two C-130s from Eglin AFB, Florida flew the lead. It is unclear whether the very secret Son Tay raid prompted the Mod 70 changes.

Since the end of the Vietnam War, the aircraft have been further modified and re-engined with 4,508-eshp (6051-kW) T56-A-15 engines used by the C-130H. In this configuration the aircraft were briefly known as C-130H(CT)s. The new designation MC-130E Combat Talon I was adopted in 1976, as a result of the Mod 70 programme. The designation replaced a variety of official and unofficial epithets, including Skyhook EC-130E and HC-130E. Twin external lights were fitted, one set filtered for use with NVGs.

During the mid-1980s, the aircraft again underwent modifications, this time for inflight refuelling, giving the MC-130E the added capability of aerial refuelling from KC-135 tankers. Extremely long-range missions were thus added to their already impressive repertoire. The MC-130E could also give fuel in flight to SOF helicopters, by means of a refuelling pod located under each wing. The flight deck is similar to that of a vanilla transport Hercules, at least insofar as the pilot and co-pilot stations are concerned. Two navigators sit side-by-side behind the co-pilot, each facing to starboard. The left-hand (forward) crewman operates the terrain-avoidance radar, while the right hand (aft) crewman operates precision ground mapping and FLIR. The EWO sits in the cargo area beside a communications specialist or radio operator. Their station is located behind the cockpit. The EWO's duties include management of the KY-58 secure communications suite, as well as the defensive systems.

Fulton today and Mod 90

Aircraft fitted with Fulton STAR now have AN/APQ-122(V)-8 radar in their reconfigured radomes. The 8th SOS continues to train using the Fulton STAR system, snatching 250-lb sandbags or rescue mannequins. The standard recovery technique involves the MC-130E-C running in at 400 ft (122 m) and 150 kt (275 km/h), aiming between red markers on the line from balloon to load. When the vertical line is snagged at the nose, it runs back under the aircraft and is snagged from the open rear ramp. The load to be recovered is then reeled in hydraulically. Deflector lines run from the nose to the wingtips when STAR is in use, to prevent a 'missed cable' from fouling the wing or props.

Another major modification programme, Mod 90, resulted in further enhancements to the avionics suite, with a WJ-1840 panoramic receiver, a digital message device group for data burst handling and 60/90-kVA generators,

Inside the MC-130E Mod 90

Above: A Mod 90 MC-130 displays its forward RWR fit. The undernose FLIR retracts behind the nosewheel door while the aircraft is on the ground.

Below: The Mod 90 cockpit is virtually identical to that of the basic MC-130E, but has terrain following radar warning lights above the main panel.

Above: Mod 90 aircraft have been fitted with an improved navigation suite including INS and GPS, in addition to upgraded radar modes and FLIR displays. The two-man nav/sensor console has been changed accordingly.

Left: The Mod 90 tailcone is bereft of the forest of RWRs carried previously. Its revised and streamlined dielectric panels obviously house more sophisticated equipment.

plus the addition of a fully integrated navigation fit with dual precision INS, new dual mission computers and an integrated global positioning system. Avionics improvements included adding a 'look into the turn' mode for the radar and a retractable integrated FLIR sensor mounted well forward, under the nose. Red and green TF warning lights are fitted above the instrument panel glare shield. A second phase covered the installation of a new electronic warfare suite with the ALQ-117 Pave Mint radar jammer, AN/ALE-40 chaff/flare dispensers and an AAR-44 IR warning receiver. This also increased the available cargo space and provided a full range of defensive countermeasures for the threats likely to be encountered in the 1990s. New RWR antenna fairings appear immediately aft of the nose radome, below the cockpit windows and above the flight deck. A square antenna below the fuselage is orientated so that the corners face fore-and-aft and at 90° to the centre-line. The tailcone is also reconfigured, with scalloped dielectric sides and a single aft-facing antenna. Structural modifications were made to the wings and wheel-well areas to extend the service life of the basic airframe, and gun-mounts were modified. The avionics are relocated forward, allowing the aircraft to carry five cargo pallets, instead of the four and a half carried by the basic Combat Talon. Photographs are not permitted inside the cargo area of Mod 90. Known Mod 90 airframes converted so far are MC-130E-Ys 63-7785 and 64-0565 and MC-130E-Cs 64-0523, 64-0559, 64-0561, 64-0562, 64-0566 and 64-0567, and all 14 will eventually receive the modifications. Nine of the surviving 14 Combat Talon Is are fitted with Fulton gear.

Combat Talon today

The type has constantly been updated and modified with new mission capabilities, special operations equipment and electronics, enhancing its mission of penetrating hostile airspace at low altitudes in poor weather or in support of night-time operations. The MC-130E Combat Talon I, along with the MC-130H Combat Talon II, are the only C-130s modified for high-speed, low-level airdrops. The Talons can drop equipment and personnel on small, unmarked drop zones with pinpoint accuracy, day or night and in poor weather. The most recent modifications will extend the service life of the Talon I well beyond the year 2000. The aircraft's primary role remains that of covert penetration of hostile airspace, to clandestinely airdrop men or material, or, in extremes, to airland behind enemy lines.

The ramp and doors are modified to allow air-dropping at normal cruising speed, both to avoid exposing the aircraft to unnecessary danger and to avoid the tell-tale slowing down which usually marks an airdrop. The advanced terrain-following radar, GPS and night-vision equipment allow extremely accurate navigation in darkness, adverse weather and at very low level (below 250 ft/75 m). Routes are heavily segmented to avoid target areas, ingress and egress points being predictable.

MC-130H Combat Talon II

The MC-130H Combat Talon II is an all-new aircraft developed to augment the Talon I. Twenty-four were delivered as green C-130H transports (with minimal equipment) for avionics installation by E-Systems with IBM Federal Systems Division handling systems integration. The Talon II has an upgraded flight deck featuring an integrated avionics mission system combining basic aircraft flight data, tactical data and mission sensor data into a comprehensive array of display formats. The pilot and co-pilot's panels are each dominated by a pair of monochrome CRT multi-function display screens, with the left screen usually used for displaying the HSI and the right displaying the ADI. The entire panel is painted light grey, rather than the black on the MC-130E, and the aircraft has redesigned control yokes with a significant number of HOTAS controls on each 'horn'.

All these upgrades greatly enhance situational awareness and crew co-ordination – vital functions during the type of low-level, night-time, adverse-weather flying that is bread and butter to these aircraft. For the first time on a Combat Talon Hercules the EWO sits on the flight deck, rather than in the cargo area, replacing one of the navigators. He sits next to the remaining navigator, facing starboard, behind the navigator and co-pilot. The navigator has two dedicated CRT display screens and a systems management keyboard, while the EWO has a keyboard and two more CRT display screens, one of which is dedicated to EW functions (mainly the ALQ-117 visual displays)and one of which can be shared with the navigator. The navigator also has dual compasses, an eight-day clock, true airspeed indicator and an altimeter. A joystick is used to operate the IDS. The EWO has an IRWR indicator, an IP-1310/ALR warning screen and a joystick to control the FLIR or radar. The flight deck and cargo areas are night-vision goggle compatible, giving the crew the flexibility of monitoring the flight controls and keeping an eye outside the cockpit simultaneously. The cargo area is particularly uncluttered with two

At the heart of the MC-130H Combat Talon II is its AN/APQ-170 multi-mode radar. It has been developed by the St Louis-based Electronics & Space Corp., solely for AFSOC's latest version of the Combat Talon. The earlier AN/APQ-122(V)8, as fitted to the MC-130E, incorporates a terrain-following mode as a secondary function to the basic C-130E radar. APQ-170 uses a far more sophisticated processor and antenna that allow the radar to fly the aircraft lower, faster and with greater safety than ever before.

Inside the MC-130H Combat Talon II

Above: The MC-130H combines APQ-170 TF radar with the Texas Instruments AN/AAQ-45 IR detection system originally developed for the HH-60 Night Hawk.

Above: Combat Talon II is fitted with a retractable AN/AAQ-15 FLIR, – a limited, but stealthy, alternative to radar for covert navigation.

Above: The navigator and EWO sit at the Integrated Control and Display Subsystem console. The nav has two displays with a keyboard, while the EWO also has a keyboard, a dedicated EW display and a supplementary nav display.

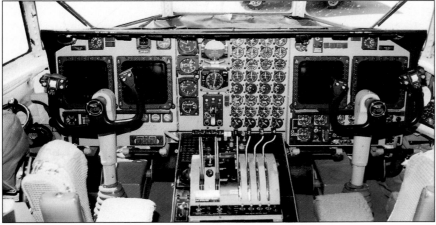

Above: The flight deck boasts four Honeywell AN/ASQ-204 Horizontal Situation Displays (HSDs) which can carry nav data, radar imagery, flight plans, and fuel and systems management information.

Above right: The MC-130H cargo bay is uncluttered by crew stations or avionics stacks on the floor.

The tailcone fairing on the Combat Talon II is surprisingly straightforward and uncomplicated in appearance.

Two twin sets of EW antennas on the fin tip are orientated at 90° to give all-round coverage. Information is transmitted to the electronic warfare officer.

aircrew relocated to the flight deck and with avionics 'cheeks' racks on the walls rather than on the floor.

The bulbous nose of the MC-130H houses the Electronics and Space Corporation AN/APQ-170 multi-mode radar which provides redundant terrain-following/terrain-avoidance, weather detection and ground-mapping capabilities, and has a 'look into the turn' capability allowing the aircraft to follow terrain contours even during a turn. False terrain images generated by the radar created unnecessary fly-up commands during early flight trials, but such teething troubles were eventually overcome. The AN/AAQ-15 infra-red detection set (IDS) can be slewed to flight path vectors and helps the pilots visually clear terrain and avoid threats. The FLIR provides a high quality image of terrain features and, combined with the radar map and other systems on board, gives the crew an accurate picture of the flight path and approaching terrain features.

The mission computer, which helps the aircraft to stay low and fast at night, also helps the crew to accurately drop supplies with reliable precision. The load and ballistics parameters are factored into the mission computer prior to take-off and are updated by the navigation systems once in the air. The pilots use the radar and FLIR to designate the drop zone and, once co-ordinates in the computer line up, the load is automatically catapulted out of the aircraft by the new high-speed low-level aerial delivery and container release system.

To enhance survivability, the Combat Talon II is fitted with an AN/AAR-44 missile launch warning system, an AN/ALQ-172 infra-red detector/jammer, AN/ALR-69 radar warning receiver and AN/ALQ-8 ECM pods. IR jammers and chaff/flare dispensers are also carried. The MC-130H Talon II can be refuelled from KC-135 and KC-10 aircraft, giving it unprecedented range. It can, in turn, be used to refuel SOF helicopters in flight using underwing HDUs. It seems odd that AFSOC C-130s are fitted with USAF-style refuelling receptacles and not probes, since this means that they cannot be refuelled by other AFSOC aircraft, while an increasing number of USAF KC-10s and KC-135s carry (detachable) drogues as well as flying booms.

The first MC-130Hs were delivered in the 'lizard' scheme but this Edwards AFB-based aircraft, detached to the 418th TS, 412th TW, has adopted the new two-tone grey scheme slowly being applied to AFSOC's C-130s.

The first four production MC-130H Combat Talon IIs, after numerous delays in the modification programme, joined AFSOC inventory in ceremonies at Hurlburt Field on 17 October 1991. The four active-duty squadrons of MC-130E/H Combat Talon aircraft are used to perform long-range insertion, extraction or resupply missions deep within hostile territory. They can also conduct PSYOP leaflet drops and deliver 15,000-lb (6805-kg) BLU-82 bombs. The MC-130H serves with the 15th SOS at Hurlburt (14 aircraft) and with the 7th SOS at Mildenhall, UK (five aircraft), while the 1st SOS at Kadena and the 8th SOS at Hurlburt use the older MC-130E. Five MC-130Hs are in service with the 58th SOW at Kirtland, used by the 550th SOS in the training role.

HC-130 Combat Shadow

Combat Shadow Hercules are the unsung heroes of the AFSOC mission. They fly much the same mission profiles as other AFSOC fixed-wing aircraft but perform the thankless and less glamorous task (unless you're a receiver begging for gas) of night-time aerial refuelling of SOF helicopters. The reduced-threat, night-time environment is rendered even safer by use of low-level tactics and by avoiding the use of lights, radios and radar wherever possible. The three active and one reserve squadrons of HC-130N/P Combat Shadow aircraft refuel SOF helicopters and support limited insertion, extraction or resupply missions.

The Combat Shadow programme name covers two tanker versions of the basic Hercules, HC-130N and the HC-130P, both originally designed for the rescue/recovery role. A handful remain in service with ANG rescue squadrons, but most have since been specifically upgraded and equipped for special operations missions. The two types are extremely similar, with the HC-130P differing in that it was originally fitted with Fulton recovery gear (which has

now been removed). The two types differed from the original HC-130H in being equipped to refuel helicopters in flight, using underwing HDUs. In AFSOC service the variants are regarded as being a common type and are generically referred to as Combat Shadows (or occasionally as HC-130N/Ps). The aircraft already have upgraded navigation, communications and threat warning and countermeasures systems, but are being further upgraded under the SOFI (Special Operations Forces Improvement) programme. With a crew of eight (pilot, co-pilot, two navigators, flight engineer, communications systems operator and two loadmasters), the Combat Shadows fly with enough onboard fuel to loiter in the target area in support of the helicopters.

Partnering the 'Giants'

Originally ordered in 1963 and first flown in 1964, the HC-130 has performed many roles and missions. In Southeast Asia they were used to refuel 'Jolly' and 'Super Jolly Green Giant' helicopters and were airborne command and control platforms for direct rescue efforts. Four aircraft were modified to launch RPVs. Fifteen HC-130Ns were built, all of which survive, although four are still used by an AFRes rescue unit. Twenty HC-130Ps were built, two of which were destroyed by satchel bombs on the ground in Vietnam. Another was lost in 1986 in a flying accident. Four serve with two ANG rescue outfits, but the remaining 13 are in the special operations community. Four more HC-130Ns were converted from HC-130Hs, these serving with special operations or special operations training units. All remaining HC-130Hs have been brought up to HC-130N standards, with inflight-refuelling capability, giving a total USAF force of 54 aircraft, of which 28 are said to be SOCOM-assigned. It is unclear as to whether this includes the aircraft assigned to the 5th SOS (an AFRes special operations unit) and to the 58th SOW (the AETC training wing). Sources suggest that the 28 Special Combat Shadows are divided as follows: 11 with the 9th SOS at Hurlburt, four with the 67th SOS at Mildenhall, five with the 17th SOS at Kadena, five with the 5th SOS at Duke Field and three with the 58th SOW at Kirtland.

HC-130 SOFI

In 1986 the active-duty HC-130Ns and HC-130Ps transitioned to the special operations mission, leaving the HC-130H to serve on in the rescue role. Remaining HC-130s are assigned to active-duty and reservist ACC and PACAF units flying purely in the CSAR role. Under the SOFI programme (which is quite separate and distinct from the similarly-named SOFI programme applied to the AC-130H), the Combat Shadows are being modified with self-contained navigation systems and inertial guidance systems, global positioning system, FLIR, radar and missile warning receivers, chaff and flare dispensers, NVG-compatible interior and exterior lighting, a head-up display and inflight-refuelling receptacles. The aircraft will also have satellite comms equipment and a data-burst device, allowing the rapid transmission of bursts of encrypted electronic data or compressed secure voice messages.

In fact, many of the HC-130s already have some, if not all of the planned modifications, which, it is said, make each 'virtually a poor man's Combat Talon!' The rescue Hercules were all originally fitted with a prominent dorsal radome above the forward fuselage. This contained the antenna of the AN/ARD-17 Cook Aerial Tracker, originally provided to help locate manned space capsules during re-entry, but later adapted to pick up signals from the locator beacons of downed aircrew. This equipment eventually fell into disuse and was removed, although aircraft retained their dorsal radomes. When the SOFI programme was instituted it was decided that the radome should be removed. On many of the first aircraft to be converted severe corrosion was discovered, and later aircraft undergoing the SOFI modification may keep their empty radomes.

MC-130Hs are now a regular sight at RAF Mildenhall. As part of the 352nd Special Operations Group, the 7th SOS exchanged its MC-130Es for MC-130Hs when the unit moved from Frankfurt/Rhein Main to RAF Alconbury. In recent times they have played an important part in Operation Deny Flight.

Perhaps the best-known AFSOC aircraft, and certainly the most glamorous, is the Hercules gunship, callsign SPECTRE. It also has arguably the longest pedigree. By early 1965, initial operations with Douglas AC-47 'Spooky' had proven the gunship concept beyond all doubt, and a larger, more capable platform was sought, the Hercules being the obvious choice. Approval of Project Gunship II was quickly granted, and Major Ron Terry and Major James Wolverton soon had the aircraft modification shops at ASD gutting former JC-130A (54-1626). By the summer of 1967, this first AC-130A, armed with four 7.62-mm Miniguns above the port undercarriage sponson and with four 20-mm M-61 Vulcan cannon fore and aft, began its flight tests on Eglin's live fire ranges. In September 1967, seven days after arriving at Nha Trang Air Base, Major Terry piloted the first AC-130A combat sortie, supporting 'troops in contact' at a firebase in South Vietnam. On 9 November 1967, SPECTRE flew its first armed reconnaissance mission over the Ho Chi Minh Trail in Laos. Within 15 minutes of arriving over the trail, six trucks were burning and SPECTRE had found its niche. By the end of the Vietnam conflict, SPECTRE had destroyed over 10,000 troop and supply-carrying vehicles along the Ho Chi Minh Trail. The next seven AC-130As were all converted from Roman-nosed JC-130As and featured improved systems, including AN/APQ-136 MTI radar, AN/AAD-4 FLIR and an AN/AWG-13 analog computer. The radar was adapted from the F-104's NASARR F-151-A. They also featured a steerable 1.5-million candlepower AN/AVQ-8 searchlight, with two Xenon arc lights, on the rear ramp.

Gunship genesis

The next AC-130A (converted from a C-130A transport) procured by the Gunship System Program Office at Wright-Patterson took on a new look, and existing AC-130As were soon converted to the same standards. Project Surprise Package replaced the two aft 20-mm Vulcan cannon and Miniguns with two 40-mm Bofors cannon. Surprise Package aircraft were also outfitted with a new AN/ASQ-145 low-light-level television system in the AN/ASQ-24A stabilised

tracking set, an AN/AVQ-18 laser target designator, a beacon tracking AN/APQ-133 radar and a new digital fire control computer. The stabilised tracking set is mounted in what was the crew entry door on a standard C-130, and can be covered by a door on the ground or during ferry flights.

Under the Pave Pronto programme the aircraft, with nine more C-130A conversions, also received the AN/ASD-5 Black Crow truck ignition sensor that had originally been intended for Surprise Package, plus provision for SUU-42 flare pods and for dual AN/ALQ-87 ECM pods. The Black Crow sensor was housed in a radome-like fairing on the port side of the forward fuselage, just ahead of the stabilised tracking set, and at its base was a perforated air deflector, which opened to counteract the yawing moment when the guns were fired.

From AC-130A to AC-130E

By 1970, structural limitations with the A-model C-130 airframes were restricting the growth of C-130 gunships. A new aircraft was needed. The Air force pulled 11 C-130E aircraft (69-6567 through 69-6577) off the Lockheed assembly line and initiated the Pave Spectre programme. The aircraft were originally intended to be similar to the Pave Pronto AC-130As, but using the improved C-130E airframe. By the time they were converted, however, they had already received a considerably improved equipment fit. Still armed with two 20-mm Vulcan cannon, two 7.62-mm Miniguns and two 40-mm Bofors cannon, Pave Spectre AC-130Es arrived at Ubon Royal Thai Air Base, Thailand in October 1971 and went to work immediately. By late 1971, North Vietnamese AAA weapons were wreaking havoc on attack aircraft operating along the trail.

The Air Force asked ASD to develop a weapon which would allow AC-130s to stand off outside the range of the AAA, but still retain the accuracy needed for truck busting. The Gunship SPO at Wright-Patterson researched a number of candidate heavy weapons for the gunship, finally settling on the trusty old US Army 105-mm howitzer in place of the rearmost 40-mm gun, thereby creating the Pave Aegis armament configuration. On 31 March 1972, a Pave Spectre

The various MC-130E sub-types were noted for the subtleties between the variants. There can be no such confusion with the MC-130H which ranks alongside the EC-130(RR) as the most obviously modified of any of AFSOC's C-130s. This is an aircraft from the 15th SOS – based stateside, it is the premier Combat Talon II unit. It is easy to forget that, despite their sophistication, the task of any MC-130 is to deliver cargo and personnel, albeit under exceptional conditions.

US Air Force Special Operations Command

Powerplant
The Combat Talon II is powered by four Allison T56-A-15 turboprops, each developing 4,508 shp (3362 kW). Each engine drives a Hamilton Standard 54H60 four-bladed, constant-speed propeller.

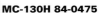

Fuel tanks
Six integral wing tanks have a capacity of 6,960 US gal (26344 litres), to which are usually added two 1,360-US gal (5146-litre) underwing tanks. The MC-130H can also carry hose-drum units for refuelling SF helicopters in flight.

MC-130H 84-0475
This aircraft was the 5041st Hercules to roll off the line and was the second, of 29, Combat Talon IIs ordered by the USAF. It was delivered in August 1985.

AC-130E modified with the new 105-mm cannon it its No. 6 gun position came upon a target which could not be knocked out by the 40-mm gun; using the howitzer, it was destroyed in short order. The 105-mm gun was initially bolted to a fixed mount, but was later fitted to a trainable mount controlled by the computer sighting system. In late 1972, the Air Force initiated a programme to upgrade the engines on their 10 surviving AC-130E aircraft. Pave Spectre AC-130Es were some of the first in line to receive the new Allison T-56A-15 engines, which, upon installation, resulted in redesignation of the aircraft as an AC-130H. The first H-model gunship arrived at Ubon in March 1973, just in time for the wind-down of US involvement in Southeast Asia.

AC-130H Spectre

The key to the AC-130H's success lay as much with the awkward-looking jumble of lenses mounted in the old crew entry door as with its firepower. The ASQ-24A stabilised tracking set might look untidy, but it is extremely effective. The cluster of lenses serves a gated laser intensifier night TV (GLINT) which increases light levels, a narrow-angle TV, AVQ-19 laser and an LTD/R (laser target designator/ranger).

By the autumn of 1975, all AC-130A and AC-130Hs left Thailand and returned to the United States, settling in northwest Florida. The 10 surviving H-model gunships remained on active duty and were assigned to the 1st Special Operations Wing at Hurlburt Field, while the 10 A models were assigned to the Air Force Reserve's 919th Special Operations Group at Duke Field. Although the gunship's day of glory was over, Vietnam proved a critical arena for the development of tactics and techniques, and of a certain esprit de corps.

IFR and SOFI for the AC-130H

In the late 1970s, the Air Force initiated a programme to outfit special operations' AC-130H and MC-130E Combat Talon aircraft, along with other selected special mission C-130s such as the EC-130E ABCCC (Airborne Battlefield Command and Control Center) aircraft with inflight-refuelling capability. Addition of the Universal Aerial Refueling Receptacle Slipway Installation greatly enhanced the AC-130H's range and time on station. In 1979, two 1st SOW AC-130H gunships set a world record for the longest-duration C-130 flight, flying non-stop from Hurlburt Field to Andersen AFB, Guam, refuelling four times in flight and landing almost 30 hours after take-off. During Operations Urgent Fury, Just Cause, Desert Storm and many others, the ability to 'top off the tanks' just prior to entering enemy airspace has afforded the Spectre the unique ability to support ground troops without interruption for up to five hours. Since the early 1980s, AC-130H combat sortie lengths in excess of 10-15 hours have been commonplace.

In December 1986, the Air Force's Warner-Robins Air Logistics Center awarded to Lockheed Aircraft Services Company, Ontario, California, a contract to modify the 1st

SOW's fleet of AC-130H gunships with a Special Operations Force Improvements (SOFI) package. This was distinct from the similarly named SOFI programme applied to the HC-130s and was aimed at improving reliability and maintainability by replacing core avionics units, computers, INS, radar and display units. The package was similar to the upgrade applied to MC-130Es under the Mod. 90 programme. In 1987, the air force expanded the scope of the modification contract to consolidate other requirements, in order to minimise aircraft down-time. Added sub-systems included a new HUD, video display systems, improved gun mounts, a new secure communications system, new EW systems, a new digital air data computer, a higher-capacity generator system, GPS and a new FLIR, dual redundant fire control computers and an improved infra-red targeting sensor. Externally, SOFI aircraft can be recognised by the relocation of their FLIR turret from the port undercarriage sponson to a position below the nose radome, and by a profusion of new antennas above the forward fuselage.

SOFI in service

The first AC-130H SOFI aircraft, 69-6568, began its transformation in December 1987, flying in its new guise on the scheduled date of 1 September 1989. It was delivered back to the 1st SOW on 31 July 1990 after extensive testing by crews from the 16th SOS and SMOTEC (Special Missions Operational Test and Evaluation Center). The ninth and final AC-130H SOFI was delivered to AFSOC ahead of schedule in August 1993. To date, AC-130H SOFI have flown hundreds of combat sorties in support of Operations Gable Shark (Desert Storm-related), Restore Hope in Somalia, Deny Flight in Bosnia, and Uphold Democracy in Haiti. One was lost in Somalia.

The SOFI programme has improved system reliability from between 50 and 100 hours MTBF to 1,500 to 2,000 hours MTBF, while raising availability to above 90 per cent. Gunfire accuracy has improved dramatically and has reached almost the standard of the new-generation AC-130U.

The AC-130H has been improved steadily, by the addition of various new and improved items of equipment. By the early 1990s, most AC-130Hs are believed to have had an AAR-44 IRWR, with a spinning sensor in a conical fairing (with a round-ended rectangular window) underfuselage, scanning the lower hemisphere for SAM launches and automatically ejecting chaff or flares. Aircraft also had the bell-shaped antenna for a Watkins-Johnson WJ-1840 microwave receiver system nearby, and a blade antenna which is believed to serve a Hallicraftres ALT-32 noise jammer. The extended tailcone houses a TRIM-7A antenna for the APR-36. The pilot has a GEC-Marconi HUD orientated to port, acting as a sight to allow him to 'aim' the side-firing weapons.

The most potent item in the AC-130H remains its crew, however. Pilot and co-pilot share the flight deck with the navigator and fire control officer, who sit side-by-side (left and right/fore and aft, respectively) facing to starboard behind the co-pilot. The fire control officer's position is

Crew
The MC-130H is flown by a basic crew of five. A typical C-130E transport relies on a flight deck crew of three – two pilots, and a navigator with a single loadmaster. The MC-130H adds an electronic warfare officer (EWO) who sits at the same console as the navigator, on the flight deck. Earlier MC-130 versions had the EWO station in the cargo area. Extra loadmasters are frequently carried.

Lockheed MC-130H Combat Talon II

This MC-130H Combat Talon II, now stationed with the 15th SOS at Hurlburt Field, was first delivered to the 6518th TS, 6510th TW, (now the 418th TS, 412th TW) at Edwards AFB, for initial Combat Talon II trials. Trials of the type were extremely protracted.

Combat controllers
Combat control teams (CCTs), and pararescuemen, are often the first AFSOC personnel inserted into a hostile area, and the last to be extracted. Both groups are trained by the 720th Special Tactics Group, at Hurlburt Field. Combat controllers act as pathfinders – securing DZs and LZs and co-ordinating communications, nav-aids and ATC for the special forces or air assault units that follow. During Desert Storm MC-130s carried CCTs deep into Iraq to help update navigational information for the coalition air forces. They also handled operations at secret FOLs and even acted as ground-based FACs.

Defensive systems
The Combat Talon II is exceptionally well protected against most threats. It carries ALQ-8 ECM pods underwing, ALR-69 radar warning receivers, ALQ-172 radar detector and jammer, APR-46 (WJ-1840) ESM, AAR-44 missile launch warning detector, QRC 84-02 IR jammer and internal chaff/fare dispensers.

Rear ramp
The ramp of the MC-130 can be opened in flight to air-drop equipment, and has baffles to allow it to be opened at higher speeds than the ramps of standard variants. This allows the aircraft to air-drop at normal cruising speed. Combat Talon II crews regularly practise high-speed, low-level air-drop tactics to deliver their load accurately with the minimum exposure to enemy air defences. Alternative methods for delivering SF parachutists are HAHO (High-Altitude High-Opening) and HALO (High-Altitude Low-Opening) drops. The first method allows the aircraft to remain miles away from the drop zone, while the second gets troops on the ground with the utmost speed.

dominated by three TV-type screens monitoring the FLIR/TV, with a computed impact display screen and a large FLIR display screen. Inside the cargo area is a 'battle management center', colloquially known as the 'booth'. This is usually occupied by two sensor operators and an EWO, but can house up to six. Observers man the bubble windows on the starboard side and on the underside of the ramp. The latter is usually the illuminator operator, and since he faces aft, has L and R markings to remind him as to which way really is right and left.

In addition to raining death and destruction from the skies, the AC-130H has various more peaceful applications, chiefly by virtue of its sophisticated night-vision systems and sensors. In 1973, for example, an AC-130H used its sensors and searchlight to locate the wreckage of a commercial airliner which came down in the Everglades, then remained on station to illuminate the crash site for rescue workers.

As part of the Congressionally-mandated revitalisation of special operations forces, a Deputy Secretary of Defense Decision Memorandum dated 22 August 1985 launched the AC-130U Gunship programme. After reviewing proposals from three contractors, on 2 July 1987 the Air Force's Aeronautical Systems Division awarded to Rockwell International of El Segundo, California, a contract to modify

a brand-new production C-130H airframe into a full-scale development AC-130U Gunship. Key design milestones of preliminary design review and critical design review were completed in May 1988 and September 1989, respectively. Based on Rockwell's firm commitment and demonstrated capability to produce the AC-130U, the Air Force awarded to Rockwell two contracts to produce 11 more AC-130U gunships, using new build C-130H-2 airframes which had less than 30 hours flying time by the time they were handed over to Rockwell for upgrade and modification.

The 'U-boat' gets underway

On 20 December 1990, design became reality as the FSD AC-130U (87-0128) lifted off the runway at Palmdale, California, landing some three hours later at Edwards AFB, California. In November 1992, the Air Force awarded to Rockwell a contract to build a 13th AC-130U – a replacement for SPIRIT 03, the AC-130H lost during Desert Storm.

From the outside, the AC-130U is significantly different from earlier gunship models, with a single trainable 25-mm Gatling gun in the forward fuselage in place of the usual twin 20-mm single-barrelled cannons. A large observer's window is installed forward of the wingroot to starboard. The U also has numerous avionics improvements, including

Lockheed AC-130U Spectre

1 Air data boom
2 Radome
3 AN/APG-80 radar scanner
4 Radar equipment racks
5 Front pressure bulkhead
6 Downward vision windows
7 Pilot's head-up-display

8 Cockpit radar repeater
9 Windscreen panels and wipers
10 Radar warning receivers
11 Overhead systems switch panels
12 Inflight-refuelling receptacle
13 Dual flight engineer's stations
14 Two-pilot cockpit with observer's seat
15 Control column
16 Rudder pedals
17 Side console panel
18 Pitot heads
19 Battery compartment
20 AN/AAQ-117 FLIR turret
21 Twin nosewheel undercarriage, forward retracting
22 Nosewheel door
23 Avionics equipment racks
24 Crew entry door and airstairs
25 Access ladder to flight deck
26 Galley unit
27 Crew closet
28 Escape hatch access ladder
29 Cockpit roof escape hatch
30 Aerial cable lead-in
31 Crew rest compartment

32 Radar warning antennas and ECM transmitter fairing
33 Flexible gun seal
34 GAU-12U 25-mm five-barrelled-rotary cannon
35 Ball Aerospace ALLTV turret

36 Canon ammunition magazine, 3,000 rounds
37 Port side equipment racks
38 Spare crew seats
39 Starboard side observer's seat
40 Observation hatch
41 UHF aerial
42 ADF loop antennas
43 Battle management centre
44 IR, TV, fire control, radar/nav and EW operators' seats and consoles
45 Hydraulic system equipment

46 Conditioned air delivery ducting
47 Wing spar attachment fuselage main frame
48 Engine fire extinguisher bottles
49 Wing centre-section integral fuel tank
50 Sigint antenna
51 Centre wing panel rib construction
52 Wing stringers
53 Starboard engine nacelles
54 External fuel tank
55 Hamilton Standard four-bladed constant-speed propellers
56 Starboard outer wing panel
57 Leading edge flush aerial panel

58 Starboard navigation light
59 Starboard aileron
60 Single-slotted flaps
61 Flap shroud ribs
62 Life raft stowage
63 Satcom antenna
64 Aileron hydraulic booster
65 Mid-cabin escape hatch
66 Emergency equipment stowage
67 Rear fuselage air ducting
68 VHF/UHF aerial
69 Fin-root fillet construction
70 ELT antenna
71 AN/ALQ-172 EW equipment packs
72 Starboard tailplane
73 Fin spar box construction
74 VOR aerial
75 Anti-collision light
76 Rudder

77 Rudder tab
78 Tail radar warning and ECM transmitter fairing
79 Rudder and elevator hinge controls
80 Elevator tab
81 Port elevator
82 Tailplane rib construction
83 Elevator hydraulic booster unit
84 Rear ramp door hydraulic jack
85 Rear escape hatch

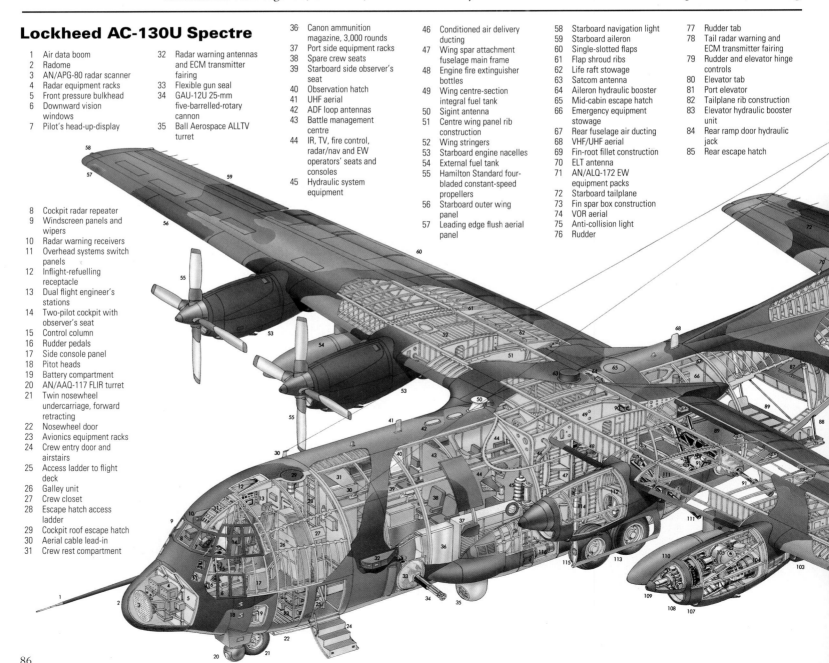

a Hughes AN/APG-180, a derivative of the F-15E's APG-70 fire control radar able to 'see' through cloud and an all-light-level TV under the forward fuselage with 360° vision. The aircraft has a Rockwell AN/APQ-172 jammer and expendable countermeasures package. With the new radar and 25-mm cannon, the AC-130U can stand off further from the target and hit targets with greater accuracy.

AC-130U problem solving

As with the MC-130H, the AC-130U suffered numerous delays in attaining particular flight test milestones, in reaching Hurlburt, and in becoming operational. A 1987 plan, announced publicly, scheduled conversions beginning in 1988, a first flight in 1989 and all 12 (later changed to 13) AC-130Us in service by late 1991. Having won the contract in competition with the well-established Lockheed (as IBM had done with the MC-130H), Rockwell found that the learning curve was steeper than anticipated. All projected dates were pushed back not once but several times.

In February 1991, a combined test force comprising flight crews, engineers and maintenance personnel from the Air Force Flight Test Center, AFSOC and Rockwell began putting the AC-130U through a tortuous series of flight tests and ground maintenance evaluations. In August 1991,

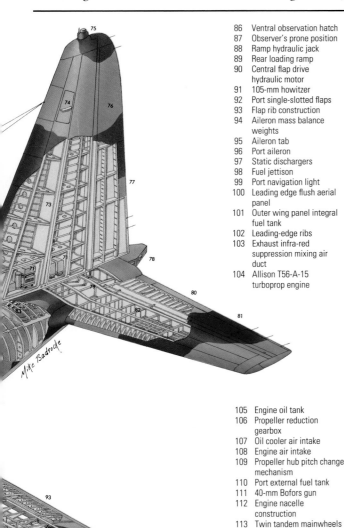

86	Ventral observation hatch
87	Observer's prone position
88	Ramp hydraulic jack
89	Rear loading ramp
90	Central flap drive hydraulic motor
91	105-mm howitzer
92	Port single-slotted flaps
93	Flap rib construction
94	Aileron mass balance weights
95	Aileron tab
96	Port aileron
97	Static dischargers
98	Fuel jettison
99	Port navigation light
100	Leading edge flush aerial panel
101	Outer wing panel integral fuel tank
102	Leading-edge ribs
103	Exhaust infra-red suppression mixing air duct
104	Allison T56-A-15 turboprop engine

105	Engine oil tank
106	Propeller reduction gearbox
107	Oil cooler air intake
108	Engine air intake
109	Propeller hub pitch change mechanism
110	Port external fuel tank
111	40-mm Bofors gun
112	Engine nacelle construction
113	Twin tandem mainwheels
114	Mainwheel leg strut and retraction screw jack
115	Landing lamp
116	Auxiliary power unit

Inside the AC-130H Spectre

Above: The AC-130H cockpit is utterly conventional, apart from the pilot's display unit (PDU) and keyboard forward of the throttles.

Left: Behind the co-pilot sits the navigator (left) and fire control officer (FCO). To the left of the FCO's panel is the FLIR monitor, and below it is the gun select panel. The central panel is the computed impact display while the 14-in monitor to the right is a second FLIR display.

Left: The AC-130H's containerised battle management centre, or 'booth', can accommodate up to six sensor operators and EWOs, but normal complement is three.

Below: Located next to the pilot's seat is the AC-130Hs GEC-Marconi HUD. This is used for weapons targetting but also has normal flight functions.

Above: The AC-130H's extended tailcone houses the TRIM-7A antenna for the APR-36 system.

The HC-130 Combat Shadow, 'the orphan of AFSOC', is, in fact, one of the Command's greatest assets. While some MC-130s can carry HDUs to refuel helicopters, the Combat Shadows are specifically intended to shepherd special operations helicopters over great distances, and have many of the MC-130s sophisticated sensors and systems. Even with auxiliary tanks, an MH-53 has an operational range of less than 700 miles. The MH-60G (as seen here) has substantially shorter legs. If SF helicopters, be they Army or Air Force, have a long way to travel, they can only do it with the help of the Combat Shadow HC-130N/P.

a combined flight crew fired the AC-130U's 40-mm and 105-mm guns for the first time, boldly pronouncing, 'This dog can hunt!'. Later that month, flight crews conducted the AC-130U's first inflight refuellings. December 1991 marked the first live fire of the new GAU-12 25-mm cannon. In December 1993, the long-awaited all-light-level television system entered the flight test programme.

On I July 1994, under the cover of darkness typical of the Spectre's way of life, a combined AFFTC/AFSOC crew delivered AFSOC's first AC-130U gunship to Hurlburt Field. Two active-duty squadrons operate the Spectre gunship (one equipped with the AC-130H and the other with the all new AC-130U) along with one Reserve squadron of AC-130A aircraft, and these provide aerial fire support or armed escort missions for both SOF and conventional forces.

Lockheed EC-130E Commando Solo

A perceived shortage of psychological warfare capability during the 1965 Dominican crisis led to the establishment of the 193rd Tactical Electronic Warfare Squadron. This unit flew EC-121S Constellations configured as airborne TV and radio broadcasting stations. These aircraft were briefly used in Vietnam during 1970 and operated elsewhere until 1979, when they were replaced by modified C-130 Hercules. Four C-130E aircraft have been modified and designated EC-130E (Rivet Rider) to support psychological warfare operations. The Commando Solo aircraft are flown solely by the 193rd SOG, which is the only Air National Guard unit with a special operations mission.

The Commando Solo normally carries a crew of 11, comprised of two pilots, navigator, electronic warfare officer, flight engineer, loadmaster and five electronic equipment

operators. While airborne, the PSYOP mission is co-ordinated by the mission control chief (the EWO) working with the five operators involved with both search and transmission duties. They control the broadcast of pre-recorded audio and video tape and the re-broadcast of signals from ground stations, and can use broadcast microphones for live transmissions. Linguists and trained broadcasters are usually provided by the US Army, and messages are prepared by the Army's 4th Psychological Operations Group at Fort Bragg. Mission transmitters include medium-frequency (MF), high-frequency (HF), very high-frequency (VHF) and ultra high-frequency (UHF). Output power is from 10 W to 10 kW. Broadcast modes include AM/FM radio, black and white and colour television, shortwave (HF) radio and other communications bands. The Commando Solo also has the capability to disrupt standard broadcasts, or to burn through or overpower existing transmissions. Mission equipment is palletised, allowing specific transmitters and receivers to be quickly installed when necessary.

The EC-130E(RR) is clearly discernible from the other variants of the C-130, although the 193rd has flown two versions of the aircraft in quite different configurations. The older model has large blade antennas on the upper fuselage adjoining the tailfin and axe-head-shaped blade antennas outboard under each wing. The aircraft have been modified to virtual C-130H standards, with inflight-refuelling receptacles and 4,508-eshp (3380-kW) T56-A-15 turboshafts.

During Operation Desert Storm three of the squadron's aircraft were modified by teams from Det. 4 of the 2762nd Logistics Squadron (Special), Lockheed Ontario and Rockwell Collins with a new TV broadcast system compatible with the format used in Kuwait and Iraq. The aircraft's long endurance and stand-off capability allowed the squadron to

broadcast safely, around the clock. During Desert Storm one crew clocked up a 21-hour mission – a C-130 record. This had a material effect on enemy morale, which crumbled swiftly, leading to the mass surrender of thousands of troops.

The latest model has a worldwide colour television broadcast system developed as a result of Gulf War experience and has four bullet-shaped VHF antennas located on an 'X' of pylons scabbed onto the tailfin sides, plus two large pods (23 ft/7.01 m in length, 6 ft/1.82 m in diameter) suspended individually under each outboard wing section. These replace smaller pods carried in the same location by the original EC-130E (RR)s. Just inboard of these are the axe-head antennas used by the original Rivet Riders. The tail-mounted VHF antennas are for lower-frequency television channels, while the steerable antennas in the wing pods are dedicated to higher-frequency television channels. Two retractable trailing-wire antennas are also fitted: a high-frequency antenna reeled horizontally several hundred feet behind the aircraft and a 1,000-ft (305-m) AM-band omni-directional antenna extended from the belly of the aircraft and held (nearly vertical) by a 500-lb (227-m) weight.

The first aircraft upgraded to the new configuration was modified by Lockheed Ontario and returned to the unit on 29 June 1994 after flutter and handling tests at Palmdale. The three remaining Rivet Riders have been similarly upgraded and may be augmented by two more conversions, although these will not be of the Comfy Levi/Senior Scout aircraft also in use with the 193rd.

Only this one Air National Guard group is equipped with the uniquely modified EC-130E(RR) Commando Solo aircraft to support psychological operations. It has a secondary mission of providing aid to the civil power during disaster relief operations, during which the aircraft can substitute for disabled ground stations, broadcasting emergency information generated on board, or acting as a flying rebroadcasting station.

Elusive Comfy Levi

The 193rd SOS also operates a quartet of EC-130E(CL)s. These aircraft are extremely secretive and have been associated with the codenames Senior Scout (which describes the mission, flown on behalf of the USAF's former Electronic Security Command), Senior Hunter (believed to refer to the two standard C-130E aircraft operated by the squadron and capable of receiving most of the Senior Scout mission equipment) and Comfy Levi. Some sources suggest that the term Commando Solo applies to these aircraft as well. For many years the aircraft and their mission were hidden behind an almost impenetrable screen of secrecy, although rumour and speculation abounded. Finally, at the 1993 Paris Air Salon, Lockheed showed a model labelled as a 'Senior Scout' and described as a Sigint (signals intelligence) platform with a 'slide-in/slide-out' equipment package.

The aircraft can look little different to a standard C-130 transport, but have a variety of antennas which can be fitted to escape hatches and other easily removable panels, including the main undercarriage doors. The aircraft are now broadly equivalent to the C-130H, with 4,508-eshp (3380-kW) T56-A-15 engines and inflight-refuelling receptacles. The 193rd SOS provides flight crews for the aircraft, but 'back-enders' reportedly come from Electronic Security Command, or from the CSS or NSA at Fort Meade. The Central Security Service at Fort Meade is responsible for capturing and recording signals, and is made up of the old Army Intelligence and Security Command, the Naval Security Group and the Air Force Security Service. The director of the CSS is also director of the co-located NSA, which has the responsibility of interpreting intercepted signals. Some reports suggest that the aircraft go to Andrews AFB for equipment installation by Fort Meade's specialists before every mission and that aircraft are de-modded here afterwards.

It was once speculated that the apparent availability of Comfy Levi airframes for conversion to the new worldwide

colour TV configuration could be partially explained by the sudden appearance of Senior Scout aircraft with a number of active-duty and ANG Hercules airlift squadrons. Such aircraft have included C-130H 74-2134 of the 463rd AW and 89-1185 of the Tennessee Guard's 105th AS. It was thought that this may indicate that these aircraft would eventually join the 193rd, or that the Senior Scout mission will be passed on to other units. In fact, extra Rivet Riders will be produced by the conversion of C-130H transports and the 193rd will retain both its Senior Hunter aircraft and the Senior Scout mission. The additional Senior Scout aircraft serving with other units are just that, extras augmenting the hard-pressed Guardsmen who claim to be the USAF's most-deployed squadron. The 193rd machines are the only inflight-refuellable aircraft which can fly the mission.

Sikorsky MH-53J Pave Low

While the C-130 can offer a degree of insertion/recovery capability through being able to para-drop or by being able to land at austere forward strips (perhaps even behind enemy lines), the helicopter often offers greater flexibility in the special operations role. The MH-53J Enhanced Pave Low III heavy-lift helicopter is the largest and most powerful helicopter in the US Air Force inventory, and is one of the most technologically advanced helicopters in the world.

Although the Enhanced Pave Low III is noted for robustness, its highly sophisticated navigation and night-flying avionics are what put it in a class by itself. The FLIR and the terrain-following/terrain-avoiding (TF/TA) radar are fully integrated with the onboard mission computer. Combined

Above: 53-3129 was built as a C-130A and was delivered to the USAF in February 1957 – the first production Hercules. It suffered a fire after its third flight and lost a wing. Subsequently it was converted to JC-130A standard and gained a large tracking radar for missile trials during the 1960s. In 1970 it was converted to AC-130A Spectre standard. Not the first of the Hercules gunships, it was subsequently christened First Lady, reflecting its status as the oldest survivor and first production aircraft.

Top: The Spectre is well protected from enemy fire but is nonetheless very vulnerable in all but the most unsophisticated theatres. This was underlined by the loss of AC-130H SPIRIT 03 during the battle for Al Khafji, which was shot down by a vintage SA-7 shoulder-launched SAM.

US Air Force Special Operations Command

Top right: Six of the USAF's 11 AC-130As remained in service with the 919th SOW's 711th SOS, at Duke Field Florida during 1995, but their days were numbered. The others languish in the sun at Davis-Monthan, their flying careers at an end. Developed under Project Gunboat, the AC-130A actually beat the AC-119G into service over Vietnam, and since then the type has undergone many transformations. 1995's AC-130As were all Pave Pronto Plus-standard aircraft, and were the most heavily armed of all the Hercules gunships, in terms of the number of guns carried. They boasted an array of two MXU-470 7.62-mm Miniguns, two M61A1 20-mm cannon and two L-60 40-mm Bofors cannon, but lacked many of the more sophisticated targeting systems carried by later Spectres, and did not carry the 105-mm howitzer. Plans to replace the AFRes AC-130As with AC-130Hs were abandoned, and the AC-130Hs and AC-130Us will remain in the active inventory.

with the ring laser gyro, inertial navigation system, global positioning system and Doppler navigation system, the Enhanced Pave Low III is capable of flying extremely low-level missions, at night, to arrive at a precise location at a specific time.

The three active-duty squadrons of MH-53J Enhanced Pave Low III helicopters provide heavy lift capabilities and are involved with medium- to long-range insertion, extraction or resupply missions in hostile territory. Its mission is transporting special operations teams to and from landing zones far behind enemy lines. Air refuellable and ruggedly built, the MH-53J can transport 37 troops and/or supplies to reinforce troops already on the ground. With its FLIR and terrain-following and terrain-avoidance radar and a projected map display, MH-53J pilots can fly ground-hugging missions regardless the terrain, weather or light conditions. In fact, under cover of darkness and with the aid of night-vision goggles, Enhanced Pave Low IIIs excel in transporting their customers on clandestine missions to landing zones deep within hostile territory. Heavily armed with either 7.62-mm Miniguns or .50-calibre machine-guns, or a combination of both, in extremes the MH-53J crews can fight their way into and out of the target area or provide fire support for special operations forces.

Pave Low lineage

The Enhanced Pave Low III airframe is not new. It is a heavily modified version of the HH-53C 'Super Jolly' helicopter used extensively in Southeast Asia for special operations and combat recovery rescue operations. Under the Air Force's Pave Low I programme, a single HH-53B was modified with low-light-level television (LLLTV), primarily for night SAR rather than for special operations tasks. An improved package was designed under the designation Pave Low II and this was fitted to a single modified HH-53B, the YHH-53H. The success of this equipment led to further refinement and modification and the YHH-53H became the Pave Low III. Eight HH-53Cs and two CH-53Cs were brought up to the same standards as HH-53Hs, with a Texas Instruments AN/AAQ-10 FLIR and AN/APQ-158 terrain-following radar, a Canadian Marconi Doppler, a Litton INS, a computer-driven projected map display, RHAWS, and chaff/flare dispensers. Extra range can be obtained by carrying a 600-US gal (2270-litre) bladder tank in the cargo hold. The surviving HH-53Hs were redesignated as MH-53Hs during 1986, after they were upgraded under the Constant Green programme, becoming the first USAF helicopters to be fully cleared to operate with NVGs.

Enhanced Pave Low

The MH-53J Enhanced Pave Low III had further improvements, including integrated digital avionics (including INS), improved radar, FLIR and GPS, a fully NVG-compatible instrument panel, improved secure communications, and internal fuel bladders, together with more powerful 4,380-shp 3285-kW) T64-GE415 engines and an uprated transmission. The Enhanced Pave Low III was produced by conversion of the YHH-53H, MH-53Hs and HH-53Cs (and seven CH-53Cs). These were modified for night and adverse-weather operations and designated MH-53J. The Enhanced Pave Low III was first funded on 10 January 1986, in an Air Staff Programme Management Document which directed that 10 (later 11) aircraft should be upgraded at a cost of $59.6 million. The prototype arrived at Pensacola Naval Air Rework Facility on 4 August 1986. Twenty-eight MH-53Js in the inventory in 1994 were due to have been joined by 13 more conversions for an eventual establishment of 41 aircraft. Many of the systems on the Enhanced Pave Low III incorporate off-the-shelf technology. The project map display and radar are from the A-7, while the ring laser gyro INS was originally developed for use with the Space Shuttle programme. The MH-53J is equipped with armour plating and a three-gun combination

of 7.62-mm Miniguns or 0.50-calibre machine-guns. It can transport 16 litters and has an external cargo hook capacity of 20,000 lb (9072 kg). The six man crew consists of two pilots, two navigators and two PJs.

In addition to the sophisticated terrain-avoidance avionics, the Enhanced Pave Low III incorporates a hover coupler system which helps stabilise the helicopter during hover manoeuvres. The hover coupler system is typically engaged a few miles prior to the landing zone and the system processes signals from gyroscopes, a radar altimeter and the helicopter's inertial guidance system to maintain a steady hover. Almost like an automatic pilot system, the hover coupler system guides the aircraft to the landing zone and allows the pilots to hold a preselected altitude above a pre-programmed point on the ground.

Although the MH-53J can deploy over extended distances using inflight refuelling, it cannot do so at jet-type cruising speeds and the logistics involved in providing tanker support for very long ferry flights can sometimes be different. The aircraft is not, however, easily air-portable aboard aircraft like the C-130 or C-141, and needs a considerable degree of disassembly even to fit into the voluminous C-5 Galaxy.

Sikorsky MH-60G Pave Hawk

Due to its small size, the MH-60G is the only true rapidly deployable, long-range special operations helicopter in the US Air Force inventory. Up to four fully configured Pave Hawks can be loaded in one C-5, where they can be transported anywhere in the world on moments notice.

The Pave Hawk's mission is to provide rapidly-deployable, worldwide, multi-mission and combat rescue capability for wartime special operations as well humanitarian assistance

Below right: Self-defence for Spectres. This AC-130A is carrying a grey QRC-84-02A IR jammer on its outboard pylon, with a green AN/ALQ-119(V)-15 ECM pod below it. Northrop's QRC-84-02 is a second-generation IRCM system. It relies on its own ram air turbine to operate independently of the AC-130A's electrical system (hence the QRC – Quick Reaction Capability – designation). The Westinghouse-built ALQ-119 pod is a 1970's design which in now in its (V)-15 incarnation. It provides multi-band (E/F, G/H and I bands) protection against hostile radars.

in peacetime. It is used to infiltrate, resupply and exfiltrate US and allied special operations forces during long-range, low-level penetrations of hostile territory at night. Because it can be refuelled in flight, the Pave Hawk has an unlimited range restricted only by the endurance of the crew. With its extremely precise navigation equipment, the Pave Hawk is the ideal platform for reconnaissance and surveillance missions where small special forces teams must go long distances to do their work.

The Pave Hawk is a twin-engined, medium-lift heli-copter optimised for long-distance flights at night and in adverse weather. As a conversion of the basic UH-60A or UH-60L it is powered by two 1,543-shp (1151-kW) General Electric T700GE700 or 1,940-shp (1447-kW) T700-GE-701C engines. The retractable inflight-refuelling probe and highly accurate, triple-redundant navigation system give the Pave Hawk incredible range and the ability to go far behind enemy lines when transporting special operations teams precisely to their destinations.

Pave Hawk configuration

With a standard crew of pilot, co-pilot, flight engineer and gunner, the MH-60G can transport between eight to 10 fully armed troops, sling-load 8,000 lb (3628 kg) of cargo and fight its way into and out of landing zones. With the external stores support system (ESSS), the Pave Hawk can carry a wide assortment of armament including two 2.75-in 19-round rocket pods, two 20-mm cannon pods or two 0.50-calibre machine-guns. The standard configuration, however, consists of two door-mounted crew-served 7.62-mm Miniguns, two 0.50-calibre (12.7-mm) machine-guns or a combination of both weapons.

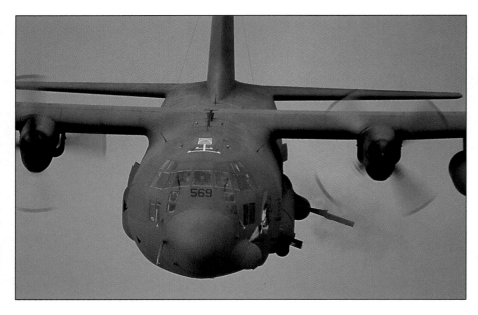

Offense is always said to be the best form of defence. The AC-130H's two 20-mm M61A1 Vulcan cannon have a rate of fire of up to 7,200 rounds per minute – and the AC-130 can carry up to 3,000 rounds per gun. The barrels have been modified, compared to those fitted in other aircraft (such as F-15, F-16 or F/A-18) to sustain prolonged bursts of fire. A prolonged burst can generate up to 4,000 lb (1814 kg) of 'sideways thrust'.

The M102 105-mm howitzer carried by the AC-130H (seen here) and the latest AC-130U is loaded manually, but aimed and fired by a computer. A good crew can generate a staggering rate of fire of eight rounds per minute. Both AC-130H and AC-130U can carry up to 100 rounds of 105-mm ammunition.

One active-duty (55th SOS) and one Reserve (71st SOS) squadron of MH-60G Pave Hawk helicopters performed short- to medium-range insertion, extraction or resupply missions, although the Reserve unit later transitioned at its Davis-Monthan base to the rescue role as the 305th RQS.

The USAF first became interested in the H-60 as a potential replacement for the HH-3E in the SAR and combat rescue roles with the squadrons of the Aerospace Rescue and Recovery Service. The original HH-60D version for the USAF proved too ambitious and expensive, and even plans for a high-low mix of HH-60Ds and more spartan HH-60Es had to be abandoned. The less well equipped HH-60A was similarly unable to attract funding, and in the end the air force had to settle for a minimum-change version of the UH-60A/L. Nineteen Credible Hawk UH-60As were delivered with or modified to have a bolt-on inflight-refuelling probe, NVG-compatible cockpits, provision for exhaust suppressors, cabin fuel tanks and folding stabilators. These served with the 55th ARRS (as it then was) and with the Kirtland-based training unit, but were soon brought up to Pave Hawk standards as MH-60Gs. The MH-60G designation was initially shared by special operations Black Hawks and rescue Black Hawks, but the latter were eventually redesignated as HH-60Gs to reflect their role and to indicate a slightly lower modification state.

Both G-model versions have a Bendix-King 1400C lightweight colour radar below the port side of the nose, an AN/ASN-137 Doppler, GPS, Carousel IVE INS and a

Teldix KG-10-20 map display unit. Both versions also have secure communications equipment, with a Motorola AN/LST-5B satellite UHF transceiver, Collins HF radios, a data burst unit and a Cubic AN/ARS-6(V) personnel locator interrogator and guidance system designed to work with Motorola AN/PRC-112 survival radios. Defensive systems include a Sanders AN/ALQ-144 IR jammer, a Dalmo Victor AN/APR-39A(V)-1 RHAWS and a Tracor AN/ALE-40 countermeasures dispenser system. The second phase of the modification programme for MH-60Gs consists of a Hughes AN/AAQ-16 FLIR in a centreline turret under the nose, Mil Std 1553B databus, a HUD projecting heading and altitude information onto the pilot's AN/PVS-6 NVGs, IR strobes for night inflight refuelling, and two new cockpit display units. The MH-60G also uses a 0.50-in M218 machine-gun in the door, rather than the usual M134 7.62-mm Minigun.

Other aircraft

The least known type in AFSOC's inventory is the CASA 212, a small number of which (believed to be four) are in service with the 427th SOS at Pope AFB, previously known as Det. 6 of HQ AFSOC. These aircraft support the US Army and joint-service special operations community at adjacent Fort Bragg and were deployed to the Persian Gulf during or immediately after Operation Desert Storm. They do not appear on the official USAF inventory and are apparently used as a low-cost method of providing jump opportunities to Army and Air Force Special Forces personnel maintaining their parachute qualifications, and for more front-line duties. A single EC-137D in VIP configuration, but equipped with satellite and other communications equipment, is used by the 2nd SOPFLT at Robins AFB, apparently for HQ support and command post duties. Unlike the Centcom EC-137Ds at the same base, the aircraft is not attached to the 19th Air Refueling Wing.

In addition to its directly assigned aircraft, AFSOC can call upon a variety of other aircraft types, including C-5 and C-141 transports assigned to Air Mobility Command squadrons and flown by specially qualified AMC crews. The 436th Airlift Wing, Dover AFB, Delaware is charged with maintaining an unknown number of C-5A/B Galaxy crews trained in low-level flying, airdrops, etc. to be available for special operations when required. The 436th would also provide any C-5s needed by special operations for transportation of equipment. There do not seem to be any special modifications to the C-5s earmarked for special operations.

Special operations C-141B

The special operations version of the C-141B lacks a distinctive aircraft designation of its own, although it does

have a distinctive package of modifications. When C-141Bs were first assigned to SOF duty (while remaining in their AMC squadrons) they had no special modifications and were no different from the standard 'trash hauler' Lockheed C-141B StarLifter.

Beginning in the late 1980s, however, an unknown number of C-141Bs (including 66-0181) were upgraded with new equipment for special operations duties. It is believed that all of these were assigned to the 437th Airlift Wing at Charleston AFB, South Carolina. Not much unclassified information is available about these aircraft, which act as an essential adjunct to AFSOC C-130 variants for operations in some environments and scenarios.

SOF C-141Bs incorporate the following modifications:
AN/AAQ-17 nose-mounted FLIR (forward-looking infra-red) set. The system is operated by a navigator. (SOF C-141Bs are the only StarLifters which employ a navigator.) The FLIR is in a non-retractable turret in the nose and can be controlled manually or automatically through interface with the radar and navigation systems. The FLIR installation is designed to provide maximum coverage throughout the complete flight regime, including climb-out and landing.

An NVG (night-vision goggle) head-up display system, or HUD. A Systems Research Laboratories Model 2745 HUD is added to the pilot's and co-pilot's positions. This displays attitude and heading reference, compass bar, vertical velocity, horizontal bar and roll/pitch ladder cage, radar altitude, flight path vector, true air speed, ground speed, true magnetic heading, bank angle and barometric altitude.

AN/AAR-44 infra-red warning receiver (IRWR). This system is integrated with the AN/ALE-40 chaff/flare dispenser. The IRWR employs a sensor in the rear fuselage to detect infra-red missile threats.

AN/ALR-69(V)-6 radar warning receiver (RWR). Indicators for this RWR are provided to both pilots and the navigator.

KY-879 data burst device. This method of secure communication is located at the navigator's station.

The modified aircraft belong to the 437th AW at Charleston AFB, SC and remain available for special operations or other low-altitude, airdrop activity. In addition, 437th AW is responsible for providing conventional (transport) support to counter-terrorist forces at Fort Bragg. The best known use of the C-141B came in October 1985 when Palestinian hijackers hijacked the Italian cruise liner *Achille Lauro*, murdering an elderly American, before seeking refuge in Egypt, where an EgyptAir Boeing 737 was provided to transport them to sanctuary in Libya. A night ambush was performed by VF-74 'Bedevilers' and VF-103 'Sluggers' aboard USS *Saratoga* (CV-60), flying

F-14A Tomcats in co-ordination with a communications intelligence RC-135 and a command-and-control E-2C Hawkeye aircraft. Seven F-14 Tomcats forced the Boeing to land at Sigonella, with the C-141B landing behind it. Aboard were American Delta Force commandos who intended to seize the hijackers and bring them to American soil; Italian guards intervened and the hijackers were prosecuted in Italy.

AMC will not reveal how many of its StarLifters or crews are earmarked for special operations and will not discuss the extent of further modifications, where modifications are carried out, how many aircraft are involved, or whether the modification programme has been completed.

AFSOC heritage

AFSOC can trace its heritage back to the establishment of the 1st Air Commando Group in March 1944 to assist British forces in Burma. During World War II special operations were undertaken on a wide scale in both Europe and the Pacific/CBI theatres, and the role was continued afterwards in the Philippines and Korea. Thereafter special operations forces were largely ignored until April 1961, when the 4400th Combat Crew Training Squadron ('Jungle Jim') was established at Hurlburt Field to train and prepare for counter-insurgency tasks. The 4400th deployed to Vietnam in November 1961 at the start of a deepening commitment to the conflict in Southeast Asia.

As the war in Vietnam expanded, the Air Force increased its counter-insurgency capability. The 4400th became a group in March 1962, and the next month became part of the newly activated US Air Force Special Air Warfare Center at Eglin. The Special Air Warfare Center obtained additional assets in the mid-1960s, including O-1 and O-2 observation planes, A-37 and A-1 attack fighters, and C-46, C-119, C-123 and later C-130 cargo aircraft, along with several types of helicopters.

In 1964, Air Commandos were deployed to Laos and Thailand on Operation Waterpump. This involved training Laotian and Thai pilots and supporting the Royal Lao army against insurgency. By 1966, Air Commandos were deployed worldwide to other countries such as Mali, Greece, Saudi Arabia, Ethiopia, Iran and the Congo.

It was the Helio Courier which came to epitomise the covert world of special operations flying. With outstanding STOL capability, a rugged structure and a capacious cabin, the Courier could land personnel and small supply loads virtually anywhere, and became the real workhorse of the clandestine insertion fleet. It was especially notable for its work in resupplying remote outposts in the Laotian jungle. In 1965 the demands for more Couriers made by the active-duty force in Southeast Asia saw ANG aircraft being

A never to be repeated formation of two 193rd SOS EC-130E(RR) Rivet Rider PSYOPS/EW aircraft. The first of these C-130s was introduced into service in August 1977. Four examples were ultimately converted to carry huge 'axe-blade' (or 'pizza cutter') antennas underwing and an even larger blade antenna on the fin leading edge, along with an entire TV studio's worth of (classified) broadcasting equipment inside. An aircraft in this configuration can be seen at the back of this formation. However, all of the 193rd SOS Rivet Riders have now been updated to provide world-wide colour broadcast capability, hence the all-new antenna fit visible on the lead aircraft.

US Air Force Special Operations Command

The 193rd Special Operations Squadron is a unique organisation – it is the only USAF unit tasked with Rivet Rider/Senior Scout PSYOP missions, and it is the most deployed USAF unit. At the same time it remains an Air National Guard squadron. The unit was established as the 193rd Tactical Electronic Warfare Squadron in 1967, most probably because the squadron (as the 140th MAS) was one of the last still flying C-121 Constellation transports into which the bulky electronic equipment could be easily accommodated. It began flying its new EC-121S' on Commando Buzz missions over Vietnam in July 1970 (though they entered service in 1968) and has retained this special mission ever since.

rushed to the regular USAF units in Vietnam. Four CONUS-based Guard units continued in the SOF task until 1975.

In 1965, the first gunships were introduced into combat with the deployment of AC-47s to Vietnam. By 1966, the high water mark for USAF special operations forces was reached with a total of 6,000 people, 550 aircraft and 19 squadrons. By 1967 the first AC-130 gunships had entered combat and the AC-119 followed in 1968. SOF operations embraced a wide variety of activities: psychological warfare with C-47s, O-2s and U-6s armed with leaflets and loud-speakers; counter-insurgency with A-1s, A-37s, A-26s, T-28s and gunships; forward air control with O-1s and O-2s; rescue support with A-1s; resupply using U-10s, UH-1s, CH-3s, C-123s and C-130s; electronic warfare with EC-47s; reconnaissance with gunships; radio relay with QU-22s; defoliation with UC-123s; plus many more specialist roles. Control of these diverse assets was largely effected by the 14th ACW/SOW at Nha Trang (later Phan Rang) and the 56th ACW/SOW at Nakhon Phanom. SOF airmen in Southeast Asia often worked closely with the 'civilians' of Air America and other CIA-backed 'airlines' in the theatre.

On 1 July 1968, the Special Air Warfare Center was redesignated US Air Force Special Operations Force (USAFSOF) and became the equivalent of a numbered air force. Perhaps more importantly, the Air Force's unconventional warriors lost the identity they treasured when their Air Commando appellation was taken away: their establishments and units were redesignated as special operations wings and squadrons. By the summer of 1968, the Vietnam War was at its peak and consumed virtually all of the Air Force's special operations attention.

The Son Tay mission

One mission mounted by USAF special operations was the Son Tay prisoner of war camp raid in November 1970. To both supporters and critics of special operations, Son Tay was perhaps the purest special operations mission ever undertaken: none of the participants were stationed in Southeast Asia, all trained thousands of miles from their destination and none had any connection with the conventional forces then located in the combat zone. Although the job of assaulting a prisoner camp to rescue American POWs was flawed by faulty intelligence – the prisoners were not there – the raid was pulled off without a single mistake or casualty and, ironically, pointed what might be possible in the future. Ironically, while Air Force leaders and many authors and observers believed that Son Tay boosted the morale of POWs, the raid had the opposite effect: POWs saw it as a sign of abject desperation.

American participation in the Vietnam War ended with the 23 January 1973 ceasefire, followed by a cessation of hostilities in Cambodia on 15 August 1973. Special operations capability declined rapidly. Most of the AC-119s and A-1s were left behind in Vietnam and newer types were trans-

ferred to the reserves or retired to the boneyard. In June 1974, USAFSOF was redesignated the 834th Tactical Composite Wing (TCW), effectively bringing to a close the long, tortured Vietnam adventure.

In July 1975, the 834th TCW was renamed the 1st Special Operations Wing (1 SOW) and by 1979 it was the only SOF wing in the air force. The Vietnam-era 14th SOW had deactivated in September 1971 to become a flying training organisation and the 56th SOW in Thailand lasted until June 1975, giving up its numberplate to a fighter wing at MacDill. The 834th comprised AC-130H Spectre gunships, MC-130E Combat Talons and CH-3E 'Jolly Green' and UH-1N 'Huey' helicopters. Two MC-130 Combat Talon squadrons remained overseas and the Air Force Reserve maintained one AC-130A gunship unit and one HH-3E 'Jolly Green' unit.

Hostage rescue

Israel's daring rescue of hostages at Entebbe in 1976 and the German operation at Mogadishu in 1977 highlighted the loss of America's special operations capability and prompted the formation of the Delta Force. When US citizens found themselves taken hostage in the US Embassy in revolutionary Iran in 1979, the weaknesses of American special operations forces were graphically highlighted. A plan was soon formulated to rescue the hostages using the Army's new Delta Force, transporting it to Tehran by Hercules and helicopter. Each of the armed services wanted to be in on the act and so USMC pilots flew USN RH-53Ds, which

were supported by USAF C-130s. This despite the fact that the USAF special operations community had pilots trained in the use of NVGs at low level over the desert, while the Marine aviators were totally unused to such flying.

Under Operation Eagle Claw three MC-130Es were to fly 139 men (truck drivers, combat controllers, Farsi-speakers and Delta Force) from Masirah in Oman to the rendezvous at Desert One. The first aircraft ('Dragon One') arrived an hour before the others, using its FLIR to check the area before landing roadblock teams and combat controllers, who set up the TACAN and landing field. The MC-130Es were to be followed by three borrowed EC-130Es (flown by 8th SOS crews) which each carried a pair of 3,000-US gal (11355-litre) fuel bladders. The EC-130Es were joined at Desert One by six of the eight RH-53D helicopters, one (BLUEBEARD SIX) having been abandoned (and its crew picked up) en route after a minor blade inspection warning and another (BLUEBEARD FIVE) returning to the Nimitz. They were to have refuelled, picked up Delta Force and flown on to a site 50 miles (80 km) further on, where Delta Force were to use locally supplied trucks to move on to the embassy and Ministry of Foreign Affairs to rescue the hostages and be picked up by the helicopters. Initially, three special C-130s were modified with take-off and braking rockets, to pick up Delta and the hostages from a soccer stadium near the embassy, but these were replaced by the RH-53Ds. At least six helicopters would be required for this phase of the operation. Unfortunately, another RH-53D had arrived at Desert One with a failed secondary

hydraulic system. With only five serviceable helicopters, the mission was cancelled, but failure turned to tragedy when one of the helicopters collided with one of the fuel-laden EC-130Es. The remaining five RH-53Ds were abandoned and all survivors were taken aboard the two C-130s.

Beyond Eagle Claw – Honey Badger

The second night of Eagle Claw would have seen four MC-130Es' worth of Rangers assaulting and capturing Manzariyeh airfield, supported by three AC-130 gunships. The hostages would have been evacuated from here aboard two C-141s and the helicopters would have been destroyed. One of the AC-130s was to have supported Delta Force in its assault on the embassy and ministry, while the other was to suppress the fighter airfield at Mehrabad. Despite the failure of Eagle Claw, a second rescue mission was immediately planned, Operation Honey Badger. This would have involved 95 helicopters and 4,000 men, including the first six HH-53Hs (transferred to special operations from the Air Rescue and Recovery Service), as well as CH-47s, UH-60s and AH/MH-6s. Two Ranger battalions were to capture Mehrabad, where the assault helicopters would be unloaded from C-5s and C-141s prior to their attacks, supported by gunships and naval fighters and attack aircraft. The operation was cancelled because the hostages were moved after the failure of Eagle Claw and proved too difficult to relocate.

In the aftermath of Eagle Claw, USAF special operations forces saw themselves transferred from TAC (which they viewed as a front-line 'warrior' command) to MAC, which

The 193rd gave up its last EC-121S in May 1979 – it was the last USAF Constellation squadron. By 1983 the squadron was a MAC-gained unit, with the Rivet Rider aircraft flying the Volant Solo mission. On 22 May 1990 the 193rd SOG/SOS became AFSOC-gained units upon the activation of this new major command, and the mission codename changed to Commando Solo to reflect this. By 1995 all four of its EC-130E(RR)s had been modified with this distinctive 'X' of fin-mounted antennas and large underwing pods, along with the 'axe-blade' antennas of the original version.

The 193rd SOS is also responsible for the shadowy EC-130E(CL) Comfy Levi Elint platform. The subtle Comfy Levi modification adds a forest of antennas to the aircraft's undercarriage doors and other (easily removable) areas such as doors and access hatches. Mission equipment is containerised and is loaded into the aircraft before every mission. When these antennas are not in place the only clue to the Comfy Levi's identity are the small 'L-shaped' antennas under the tail. The Comfy Levi flies the Senior Scout mission, which is believed to involve disruption of enemy military communications networks.

was viewed as a second-line, trash-hauling command. MAC's intention was to integrate AFSOF into its Aerospace Rescue and Recovery Service and this resulted in TAC insisting that AFSOF should retain its special identity after the transfer, with separate subordinate commands for AFSOF and ARRS. On 3 March 1983 all Air Force special operations were merged under the 2nd Air Division of the 23rd Air Force, Military Airlift Command. Major General Hugh Cox was the 2nd Air Division's first commander and helped ease the transition, but was promoted out of the AFSOF world when he made brigadier general, despite the post of vice commander 23rd Air Force falling vacant at the

same time. New personalities in 23rd Air Force began to try to absorb AFSOF into the rescue world again. Cox maintained a strong interest in special operations issues, co-operating with Hurlburt's local Congressman, Earl Hutto, and submitting various discussion documents and recommendations to whoever would read them. He believed that while non-SOF people controlled SOF funding and operations it would remain a low priority, and pushed hard for the establishment of a separate command, a quest which bore fruit in 1987 with the establishment of SOCOM.

Even with SOCOM created, there were many battles fought to maintain the separation of AFSOF and the

A number of HH-53Bs, CH-53Cs and HH-53Cs, along with eight existing MH-53Hs, were upgraded to MH-53J standard, between 1986 and 1990, under the Enhanced Pave Low III programme. It is still uncertain as to exactly how many MH-53Js have been produced. The Enhanced Pave Low III programme added capability through the addition of further night/adverse-weather avionics and systems.

ARRS. Operational squadrons were successfully separated, but MAC did manage to combine SOF and SAR training at Kirtland. An entirely separate AFSOC was created in 1990 as a major command reporting directly to SOCOM, with most of the USAF's HH-53Hs (modified to MH-53 standards) and most HC-130Ns and HC-130Ps. Some battles began to be won by the special operators. In October 1987, for example, the AFRes 71st SOS stood up to take over the aircraft and personnel of the 302nd ARS at Luke AFB, moving to Tucson to take up a drug interdiction role. The squadron traded its CH-3Es for HH-3Es in 1990 redesignating these as MH-3Es. The unit was called to active duty in December 1990 and four of its aircraft (fitted with FLIR turrets, GPS and APR-39A(V)1 RWRs) deployed to Saudi Arabia. Unofficially dubbed 'Pave Pigs', they were assigned overwater SAR duties (escorted by MH-60Gs) but also performed SOF insertions and extractions. Some equipment, including ALQ-144(V)3 IRCM jammers and Tracor AN/ALE-40 chaff/flare dispensers, did not arrive in time and crews improvised by standing in the doors with flare pistols at the ready. The M240E gun was another non-arrival, and the aircraft retained door-mounted M60Ds. The squadron transitioned to the MH-60G and continued its drug interdiction and rescue tasks, but reverted to being an air rescue squadron in March 1994.

SOCOM at war

The first trial by fire since Vietnam occurred when gunships from Hurlburt departed on the evening of 24 October 1983 to participate in Operation Urgent Fury, the invasion of Grenada. The all-night flight from Florida to Grenada took almost 10 hours and required two heavy-weight inflight refuellings.

The first AC-130H arrived over Port Salines on 25 October and, while performing area reconnaissance, saw that the airfield was littered with telephone poles, trucks and other obstructions. It was soon decided to air drop the ground forces as opposed to landing them on the now obstacle-strewn airfield.

Inside the MH-53J Pave Low III

Left: The most obvious addition to the MH-53J is its nose-mounted sensor suite which includes the AN/APQ-158 terrain-following radar and the AN/AAQ-10 FLIR turret below it. Also visible here are the forward RWR antennas and one of the six-barrelled 7.62-mm 'buzz-guns'.

Right: Texas Instruments supplies the Pave Low III FLIR, an AAQ-10. The square fairings behind the FLIR turret are for the aircraft's radar altimeter – another vital element of the MH-53J's low-level, all-weather capability.

Left: The cluttered interior of the MH-53J cabin is made even more crowded by the addition of the 600-US gal (2271-litre) bladder fuel tank. Hanging on the starboard wall is a litter for casualty evacuation.

Right: Mounted on the sponson above the fuel tank on either side is the ALQ-157 IR jammer.

Below: On the rear ramp is a 0.50-in machine gun, fitted with a chute for spent cases. The gunner crouches at the end of the ramp, wearing a monkey harness.

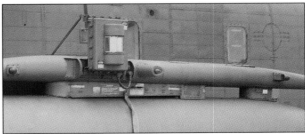

US Air Force Special Operations Command

After Desert Storm, some of the Enhanced Pave Lows flown by the 20th SOS 'Green Hornets', at Hurlburt Field, retained this two-tone desert scheme. Beneath the nose of this aircraft is the SX-5E controllable infra-red light – a 500-watt xenon spotlight that provides much needed illumination for night operations.

Mounted on the rear of this aircraft are four AN/ALE-40 chaff/flare dispensers. MH-53J crews often practise with M-208 flares over the 'live' training ranges, and blocks of flares whistle alarmingly past the head of the rear gunner when they are fired.

At daybreak the Spectres went into action against both AAA installations and ground forces. The AC-130Hs put down an accurate pattern of fire, and were instrumental in the eventual victory.

Just Cause

For Just Cause – the 1989 invasion of Panama – special operations forces were at the forefront of the action performing strike operations at Torrijos/Tocumen airport, Rio Hato airfield, La Comandancia Complex, Paitilla Airfield, Balboa Harbor, Colon, Pacora River Bridge and other sites. More than 4,400 special operations forces, including psychological operations and civil affairs forces, participated in H-hour and subsequent follow-on missions. Because of their language proficiency and cultural awareness training, SOF personnel were instrumental in achieving the surrender of entire garrisons of Panamanian soldiers, undoubtedly saving lives on both sides of the conflict and preventing enormous destruction. During Promote Liberty, USSOCOM forces played a key role in the nation-building process in Panama.

From 20 December 1989 to 7 January 1990, 23rd Air Force spearheaded Operation Just Cause. This was aimed at unseating President Manuel Noriega, the CINC of the armed forces who had effectively seized power from the legitimate elected government and who was involved in drug smuggling and gun running.

The plan in Panama called for 27 separate and simultaneous raids, airdrops or attacks against 11 different locations. Five

MH-53J Pave Lows from the 20th SOS participated in several engagements, including the catastrophic landing of Navy SEALs at Patilla airport to take out Noriega's personal Learjet. MH-53Js successfully dropped the SEAL assailants offshore at Patilla but, because of faulty intelligence, they were sorely outnumbered when they attacked the airfield and had four men killed – including the SEAL who took out the Learjet with a round from a LAW (light anti-tank weapon). A 55th SOS Pave Hawk was also called in to support the SEAL team pinned down in a fierce exchange of gunfire.

Guardsmen enter the fray

For the first time, the 193rd Special Operations Group, found itself deeply involved in conflict. During the operation, the 193rd SOG, based at Harrisburg, Pennsylvania, flew its EC-30Es to Panama and transmitted messages not only interrupting Noriega's televised broadcasts but also encouraging civilians to stay indoors and away from military installations. One mission, which included air refuelling, lasted 21 hours.

The gunship crews from Hurlburt and Duke Field obtained valuable experience over Panama and proved beyond a doubt just how effective the Spectre is in providing precision firepower. Although the shooting was over soon enough, several active-duty and Reserve gunship crews gained recognition for distinguished action. Just Cause also marked the introduction of the new multi-ship tactic known as 'Top Hat'. Only 30 days prior to the action in Panama, Major Emmett 'Otis' Redding developed this system whereby two gunships working the same airspace flew in concentric orbits.

The battle in Panama was a 'made for gunship' war – fought at close quarters, at night, where precision and lethal firepower was needed in a tightly restricted area. It provided plenty of 'firsts' for gunship crews; the first 'Top Hat' missions where two aircraft flew concentric circles firing within 50 ft (15.25 m) of each other, the first use of night-vision goggles in combat and, without a doubt, the first combat any of the crews had seen where bullets were going and coming.

AFRes battle honours

It was also the first time Air Force Reserve crews were used without being officially called up. For the 919th Special Operations Group, normally based at Duke Field, Florida (just down the road from Hurlburt Field), it was a case of being in the right place at the right time. Earlier in the month, two AC-130As of the 711th Special Operations Squadron had been despatched to Howard Air Force Base to conduct joint training exercises with US Army and Marine units and to provide security for facilities in Panama.

As the political situation between the Noriega government and the US diplomatically fell apart, the 919th SOG found themselves as the only AC-130 unit on hand when hostilities kicked off (the aircraft from Hurlburt were still en route). Although the Reserves were not involved in the invasion planning or, for that matter, in any of the operational plans,

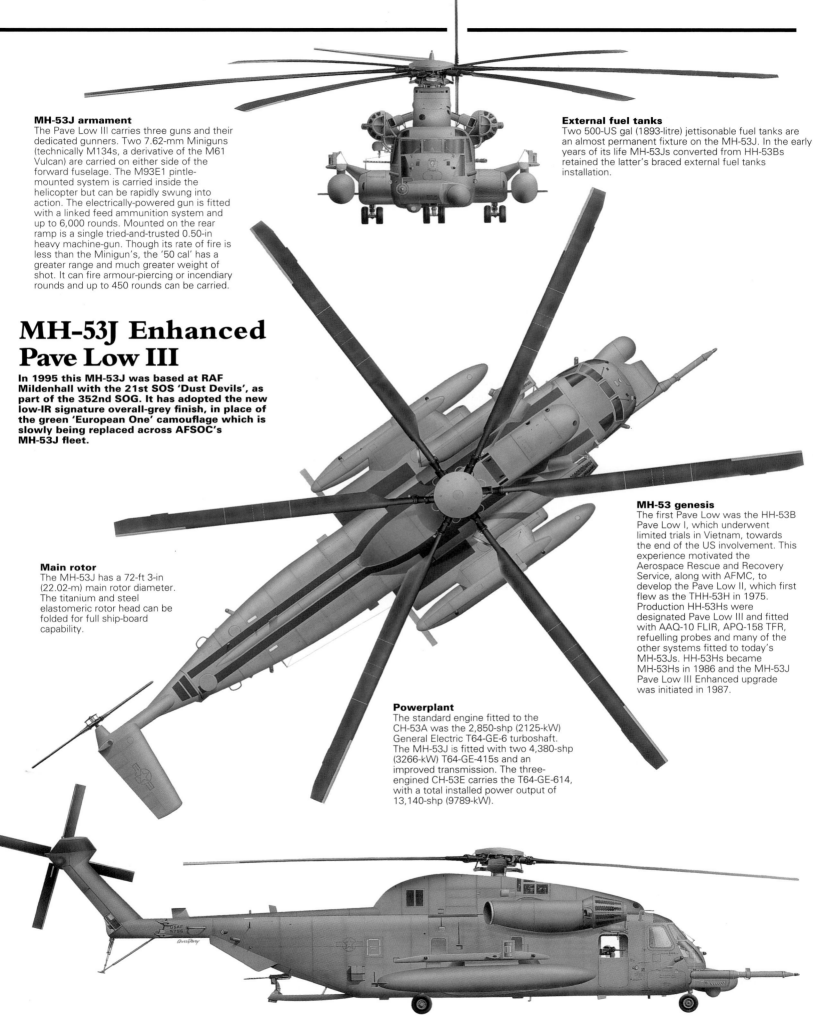

MH-53J armament
The Pave Low III carries three guns and their dedicated gunners. Two 7.62-mm Miniguns (technically M134s, a derivative of the M61 Vulcan) are carried on either side of the forward fuselage. The M93E1 pintle-mounted system is carried inside the helicopter but can be rapidly swung into action. The electrically-powered gun is fitted with a linked feed ammunition system and up to 6,000 rounds. Mounted on the rear ramp is a single tried-and-trusted 0.50-in heavy machine-gun. Though its rate of fire is less than the Minigun's, the '50 cal' has a greater range and much greater weight of shot. It can fire armour-piercing or incendiary rounds and up to 450 rounds can be carried.

MH-53J Enhanced Pave Low III

In 1995 this MH-53J was based at RAF Mildenhall with the 21st SOS 'Dust Devils', as part of the 352nd SOG. It has adopted the new low-IR signature overall-grey finish, in place of the green 'European One' camouflage which is slowly being replaced across AFSOC's MH-53J fleet.

External fuel tanks
Two 500-US gal (1893-litre) jettisonable fuel tanks are an almost permanent fixture on the MH-53J. In the early years of its life MH-53Js converted from HH-53Bs retained the latter's braced external fuel tanks installation.

Main rotor
The MH-53J has a 72-ft 3-in (22.02-m) main rotor diameter. The titanium and steel elastomeric rotor head can be folded for full ship-board capability.

MH-53 genesis
The first Pave Low was the HH-53B Pave Low I, which underwent limited trials in Vietnam, towards the end of the US involvement. This experience motivated the Aerospace Rescue and Recovery Service, along with AFMC, to develop the Pave Low II, which first flew as the THH-53H in 1975. Production HH-53Hs were designated Pave Low III and fitted with AAQ-10 FLIR, APQ-158 TFR, refuelling probes and many of the other systems fitted to today's MH-53Js. HH-53Hs became MH-53Hs in 1986 and the MH-53J Pave Low III Enhanced upgrade was initiated in 1987.

Powerplant
The standard engine fitted to the CH-53A was the 2,850-shp (2125-kW) General Electric T64-GE-6 turboshaft. The MH-53J is fitted with two 4,380-shp (3266-kW) T64-GE-415s and an improved transmission. The three-engined CH-53E carries the T64-GE-614, with a total installed power output of 13,140-shp (9789-kW).

Above: Part of the MH-53J's maximum take-off weight of 22,680 lb (50,000 kg) comprises 1,000-lb (454-kg) of additional titanium armour for the crew. Despite this, the aircraft is surprisingly manoeuvrable and has a maximum level speed (clean) of 196 mph (315 km/h).

Opposite page: Partnering the Pave Low is the diminutive MH-60G Pave Hawk. Sikorsky's MH-60G is geared around inserting and extracting small SF teams – with a standard crew of four, the Pave Hawk can carry a maximum of 10 fully armed troops, although a smaller group would be more normal.

tasking. Volant Solos broadcast for nearly 20 hours every day. During their time in the desert the 193rd also passed a significant milestone in the unit's history by amassing 130,000 hours of accident-free flying.

Throughout Desert Storm the Combat Shadows of the 9th and 67th SOS joined up with helicopters, often well into Iraqi territory, to 'pass gas' and enable the helicopters to continue their missions and return safely to base. They eventually racked up over 300 flying hours in 103 combat sorties, performing air refuelling, combat search and rescue and special operations airlift.

MC-130Es head east

The four deployed MC-130Es of the 8th SOS carried out numerous psychological operations flights as well as combat search and rescue missions. PSYOP missions consisted of leaflet drops aimed at inducing Iraqi forces to surrender, and it is estimated that more than 300,000 troops were exposed to the campaign. The weather in the theatre resulted in good 'bullshit-bombing' conditions, with little spread of the leaflet packages. The main tactic was 'operant conditioning', using PSYOP (which included ground stations, Volant Solo and leaflet drops) to inform ground troops of a particular unit that they were about to be attacked. Next day three B-52s would visit the unit, to be followed by yet more leaflets informing the survivors that the bombers would be back. This they would do, by which time many soldiers were all too willing to believe anything that PSYOP told them. No threats were left unfulfilled, a vital part of the whole PSYOP plan. Around eight B-52s were dedicated to this campaign. Approximately 29 million leaflets were delivered during Desert Storm, mostly from Combat Shadows and Talons.

Perhaps the MC-130's greatest claim to fame, though, came through its 'bombing' missions. Out of the 27 combat sorties flown by Combat Talon crews in Desert Storm, none were more unusual than the five missions in which the 15,000-lb (6804-kg) BLU-82 'Daisy Cutter' bombs were dropped. Combat Talons dropped a total of 11 BLU-82s which resulted in some of the most dramatic explosions of the war.

The standard procedure for dropping the bomb was from 6,000 ft (1828 m) but, since AAA in the target areas was intense at that altitude, tactics were modified to drop the bombs from 16,000 to 21,000 ft (4876 to 6400 m). It was further agreed to make multiple drops in order to increase the psychological impact on the enemy and to take advantage of the first bomb going off. The theory was that, after the first bomb went off and the enemy poked their heads up to see what in the world had happened, the second bomb would go off with even more devastating results.

The 'Blues Brothers'

On 3 February 1991, the first two of an eventual 18 BLU-82s arrived at King Fahd International Airport from Hill Air Force Base, Utah. During the night of 6/7 February, an MC-130E Combat Talon I dropped the first of these on a mine field in order to blast a safe passage through the Iraqi defences. The second bomb was dropped on an Iraqi battalion headquarters which resulted in the surrender of the commander, two intelligence officers and a private. The prisoners cited the bombing as an influential factor in their surrender. BLU-82 missions were usually preceded by an MC-130E leaflet drop with Arabic language leaflets stating that "Tomorrow, if you don't surrender, we're going to drop the largest conventional weapon in the world on you", and were followed with leaflets stating that "You have just been hit by the largest conventional bomb in the world. More are on the way." On 14 February a double drop occurred in the tri-border area with reportedly massive enemy casualties.

The third BLU-82 mission took place on 18 February when three MC-130Es dropped three 'Daisy Cutters' on Failaka Island. On the way in to the target, the three Talon

they were quickly notified on 19 December, only six hours before H-hour, that they were needed in a shooting war.

War in the desert

From early August 1990 to late February 1991, AFSOC participated in Operations Desert Shield and Desert Storm, the biggest conflict since Vietnam for US armed forces. AFSOC units were rapidly deployed to Saudi Arabia and to Turkey and were among the first units to arrive in theatre. In the north, the 39th SOW deployed detachments from its 7th SOS (MC-130E), its 67th SOS (HC-130) and its 21st SOS (MH-53J) to Incirlik, with a forward deployment base at Batman (inevitably nicknamed Gotham City), 150 miles (240 km) from the Iraqi border. An unusual mission performed by the Turkey-based helicopters was to transport ordnance experts into the desert to blow up bombs jettisoned by returning F-111s, to prevent the explosives from falling into the hands of the Kurds.

In southern Saudi Arabia the 193rd SOG, from Harrisburg, initially deployed two EC-130Es at the outset of Desert Shield. The Saudi Arabian authorities prevented any PSYOP activity during the early part of Desert Shield, so the first live broadcast was not made until Thanksgiving Day (22 November) 1990, when Voice of America broadcasts were made to Iraq. In January 1991, a third EC-130E deployed to a second location in the region. During the war the unit's aircrews averaged more than 14 flying hours per day for 60 consecutive days, and far exceeded their wartime

Is encountered AAA fire. As the three bombs exploded (five seconds apart) all AAA fire ceased and enemy resistance subsequently collapsed. The island fell to friendly forces within a matter of hours.

The MH-53J also proved its worth in the opening hours of Desert Storm. A plan was drawn up, codenamed Instant Thunder, where members of the 5th Special Forces Groups would attack three Iraqi radar sites just opposite the northwest border of Saudi Arabia. This attack would open a hole in the radar coverage through which coalition aircraft could fly undetected. The 3rd Battalion of the US Army's 160th SOAR was tasked with getting the special forces teams out once the radar sites were destroyed. The 20th SOS with its MH-53J Pave Lows were assigned the task of combat search and rescue for any strike aircraft going down in Iraq and for the retrieval of any special forces personnel that might become casualties in the initial assault.

Due to several contributing factors, not least of which was the lack of global positioning systems man-packs in theatre, the project was cancelled pending a more reliable plan. In the interim, the Iraqis moved the radar sites from the original 3 to 4 miles (4.8 to 6.4 km) from the border to between 20 and 40 miles (32 and 64 km). Since the radar sites had to be knocked out prior to the air assault, General Schwarzkopf gave the mission top priority. The 20th SOS Enhanced Pave Lows were found to be the only aircraft in theatre with GPS. While their MH-53s could easily reach the targets at a precise time in the almost total darkness, the lack of heavy firepower left some question as to whether the Pave Lows alone could totally knock out the Iraqi communications with their 0.50-in machine-guns.

A suggestion was made and quickly adopted to combine the attributes of the Army's AH-64 Apache and AFSOC's MH-53J Enhanced Pave Low. By combining the firepower of the Apache with the precision navigation of the Pave Low, the coalition forces now had a strike package that could do some real damage.

Task Force Normandy

The operation was codenamed Task Force Normandy and for four weeks the Apache and Pave Low crews worked together, complementing each other's strengths and perfecting their plan to take out the Iraqi 'Spoon Rest' mobile early warning radar, the 'Flat Face' early warning and target acquisition radar and the 'Squat Eye' search and target acquisition radars.

At 14.00 hours on 16 January 1991 the 20th SOS crews at Al Jouf received their orders to 'start the war'. Time over target was set for 22 minutes prior to H-hour and take off was set for 01.00 hours. Since the coalition's armada of fighter/bomber aircraft was slated to pour through the radar gap by 03.00, there was absolutely no margin for error.

At 02.12 hours the 13 aircraft of Task Force Normandy crossed the border into Iraq. Crews later reported that the night was pitch black and that the Pave Low crews had to rely totally on the computers and sensors in their cockpits as they flew no more than 50 ft (15.25 m) over the trackless desert dodging Iraqi patrols and avoiding radar detection. The previous week's intense training paid off. Despite poor visibility due to blowing sand and the disorienting effects of flying low over the desert, Pave Low crews dropped glowing chemical sticks on the ground to help guide the Apaches. As they overflew the glowing markers the Army pilots updated their navigation systems.

With only 1 per cent moon illumination, the Apache formation made visual contact with the targets almost 7.5 miles (12 km) out and achieved positive target identification at 4.3 miles (7 km). As they approached the first target, an Iraqi sentry was seen making a run for one of the bunkers. He was a second too late as the bunker exploded from the impact of an AGM-114 Hellfire missile. The two radar sites were destroyed almost simultaneously in a barrage of Hellfire missiles and Hydra 70 rockets.

up on radar but were unable to get any shots off as the high-flying F-15s interceded and chased the enemy aircraft from the scene.

Frustrated and low on fuel, the Pave Low crew returned to Al Jouf and, while in the hot pit taking on fuel with engines running, received word that an A-10 pilot had good co-ordinates for, and voice contact with, Lieutenant Jones. Encouraged by the sudden turn of events, two MH-53s immediately launched with Captain Trask and Major Homan leading the flight. Flying low to avoid detection and playing a high-stakes game of hide and seek with enemy Roland SAM sites, the Pave Lows continued on their flight path, all the while dodging enemy airfields and at least one major highway busy with considerable truck traffic. After waiting for the traffic to pass, one MH-53 held south of the highway as back-up in case they were needed as Captain Trask and his crew, now joined by the two A-10s, continued on. The A-10 pilots alerted the helicopter crew that when the formation passed over Jones' position the lead A-10 would go vertical. Only a few minutes later the lead A-10 pilot went through his climbing manoeuvre as the Pave Low crew made voice contact with Lieutenant Jones. His position was 25 miles (40 km) north of the first rescue attempt earlier in the day. As Captain Trask and Major Homan flared their helicopter for landing one of the door gunners, Master Sergeant Tim Hadrych, spotted an enemy vehicle headed directly for Jones' position. As the helicopter broke to the right, Major Homan called for one of the A-10s to 'smoke the truck'. The A-10s obliged almost instantly. As the burning remains of the enemy truck rolled crazily out of control, the Pave Low settled to the ground. Two PJs exited the aircraft and quickly helped Lieutenant Jones aboard and, in a cloud of dust, the Pave Low took off and made its way back to Al Jouf without further incident.

The crew of MOCCASIN 05 who participated in the SLATE 46 mission were awarded the 'Rescue Mission of the Year' title by the Jolly Green Association, but this did not bury the controversy surrounding the performance of AFSOC units in the CSAR role. Ironically, although Jones was plucked to safety, not a single Air Force flier shot down during Desert Storm was rescued and there have been some mutterings that, in their first outing as part of their own independent major command, AFSOC's combat rescue crews performed poorly. This allegedly further harmed the morale of those pilots and PJs who had never wanted to be special operators at all. More objective observers (such as Richard Hallion, in his book 'Storm over Iraq') felt that while CSAR forces "operated with their traditional reputation for bravery, dedication and willingness to take chances to rescue downed aircrew, there were simply too few aircraft available to meet the requirements of both CSAR and special operations needs. In many cases downed aircrew were so deep within enemy territory, or immediately surrounded by hostile forces, that rescue was impossible."

AFSOC sees out the war

Throughout Desert Storm the MH-60G Pave Hawk crews also supported Special Forces teams while working in concert with the MH-53J Pave Lows of the 20th SOS. For most of the crews it was their first taste of combat. The two helicopter types took part in virtually the closing act of the war, rappelling special forces soldiers onto the roof of the US Embassy in Kuwait City on 26 February during the recapture of the city.

Gunship crews were also called into action. From 17 January to 27 February 1991, four active-duty AC-130H gunships flew 50 sorties from a site which was still under construction, accumulating over 280 hours. Their most successful missions initially appeared to be those which involved deep penetration sorties to destroy early warning radar and known or suspected 'Scud' sites. It later transpired that the major campaign against mobile 'Scud' sites was

The warning radars were silenced immediately and the mission was a complete success. But it was not without its scary moments. En route from the targets one of the Pave Lows was engaged by two SA-7 missiles, both inside Saudi airspace. Gunners aboard the Pave Low called for the pilots to take immediate evasive action and jettison their countermeasure flares which, fortunately, caused the missiles to miss their target. Despite the near-miss, all aircraft and crews returned safely to their bases.

Tomcat rescue

Once the air war was in full swing the MH-53s were increasingly to become involved involved in the personnel recovery role. The oft repeated 'if we can find them we can get them' belief expressed by SOCOM Pave Low crews was put to the test on 21 January, 1991, in the daring rescue of downed Navy pilot Lieutenant Devon Jones. His F-14 had been shot down earlier in the day and there were good reports that both he and his backseater Lieutenant Larry Slade had survived the ejection. Two MH-53Js were scrambled at 06.00. There was no time for in-depth mission planning as the crews quickly boarded their aircraft and headed across the border. The weather was terrible. Visibility was near zero. Despite the poor flying conditions Captain Tom Trask and his co-pilot Major Mike Homan 'crossed the fence' into Iraq and pressed northward toward the co-ordinates supplied by AWACS. CAP was provided by USAF F-15s.

The first search proved futile as the only wreckage spotted in the desert was that of a Soviet-built aircraft and, upon closer examination, even that was determined to be from an old crash. As the Pave Low crew pressed their search, two Iraqi helicopters and an enemy Mirage F1 locked them

The MH-60G's four-bladed titanium/composite main rotor has Kevlar tips swept to 20° and is capable of sustaining a hit from a 23-mm shell. Up front, the Pave Hawk is fitted with a retractable inflight-refuelling probe, Bendix-King 1400C lightweight colour radar, and Hughes AN/AAQ-16 FLIR. Mounted in the door is a M134 Minigun.

The MH-60G is chiefly tasked with supporting Army Special Forces teams and this can often include missions over water. The Pave Hawk's reliable twin 1,560-shp (1151-kW) General Electric T700-GRE-701C turboshafts make hovering over water a slightly more attractive proposition, though it is a situation that no helicopter pilot would ever choose to be in.

US Air Force Special Operations Command

Above: For those who consider the aircraft of Air Force Material Command 'boring', the 437th Airlift Wing has the antidote. The Wing operates an unknown number of modified StarLifters, which undertake unknown missions as part of AMC's little-known special operations capability.

Right: The CASA Aviocars of the 427th Special Operations Squadron, based at Pope AFB, provide small 'jumpships' for special forces personnel, supposedly for training purposes. They have, however, been deployed to northern Iraq.

Top right: The Dover-based 437th Airlift Wing is responsible for C-5 special ops. Galaxies tasked with this role are unlikely to be specially equipped, or to carry any external signs of their mission and it is more likely to be a crew training qualification than a project resembling that of the 436th's C-141Bs.

Right: This Boeing EC-137D is attached to the 2nd Special Operations Flight. Its principal role is transport of SOCOM staff, and it is mintained by AMC's 19th ARW. The EC-137D designation was previously applied to the two protoypes of the E-3 Sentry.

almost totally unsuccessful, however. Initial problems encountered by the gunshippers included the intense heat on the ground, which wreaked havoc with the AC-130H's sensitive electronic equipment, and the poor visibility, with dust and sand hanging in the air and reducing the efficacy of the FLIR and LLLTV, while airborne grit scratched lenses and filters.

The battle for Al Khafji was the first opportunity in Desert Storm for gunships to directly support ground operations. Two sorties were launched on the evening of 29 January 1991 to support coalition forces engaging the Iraqis. However, the next day a 16th SOS AC-130H (callsign SPIRIT 03) was shot down in the early morning hours as it tried to

engage an Iraqi FROG missile site, having made the mistake of remaining on station half an hour after dawn. It was a terrifying reminder of the gunship's vulnerability over the modern battlefield. All 14 crew members were killed when their aircraft had its wing blown off by a surface-to-air missile and crashed into the Gulf. The 'old warriors' of the 919th who had distinguished themselves over Panama brought five of their ageing AC-130As.

Special operations aviators also saw more active service in Somalia. On 11 June three AC-130Hs operating from Djibouti attacked targets around Mogadishu and a Somali radio station being used by General Adid to broadcast propaganda. Somalia also saw a major loss for the special operations community. A 16th SOS AC-130H was lost after take-off from Mombasa, Kenya on 14 March 1993 with the loss of eight of the 14 aboard.

War in Europe

In June 1991 tensions in Yugoslavia turned to civil war as one republic after another seceded from the federation. As Slovenia, Croatia and Bosnia-Herzegovina in turn fought for their independence, so the level of violence increased, culminating in widespread fighting in troubled Bosnia. With Serb forces laying siege to the capital Sarajevo, a multi-national air bridge commenced in July 1992 to bring food to the city. Known as Operation Provide Promise, the air bridge involved regular C-130 units, although some MC-130s were employed. Meanwhile, UN road convoys attempted to reach outlying towns. Many of these road routes were cut, leaving several enclaves stranded and starving.

To allay the immediate problem, air-drop operations began on the night of 27/28 February in the face of Serbian advances on Moslem-held territory. The first mission was a

leaflet drop by MC-130Hs from the 7th SOS at Alconbury. Some 600,000 of these leaflets were intended to warn the warring factions not to fire upon successive waves of aid-carrying aircraft on subsequent nights. C-130s from the 435th AW at Rhein Main, augmented by aircraft from the 317th AW at Pope AFB, began a programme of aid drops, often with AFSOC MC-130s using their advanced navigation and night vision equipment as pathfinders. The first food drop was made from 10,000 ft over Cerska on 28 February/1 March, with three C-130s dropping 23 tons (21 tonnes), including 1.6 tons (1.5 tonnes) of medical supplies. These fell more than 0.75 mile (1.2 km) from their aiming point, although accuracy improved dramatically on subsequent missions, the second of which went to Zepa. The main food items were US MRE (Meal Ready to Eat) ration packs, compact enough for 8,000 to be carried by each Hercules.

Bosnia rescue

When the US Marines Mediterranean-based Marine Expeditionary Unit moved to Somalia, the USAF established a detachment of four MH-53Js and four HC-130N/Ps at Brindisi to make up for the loss of long-range CSAR capability. These have been manned primarily by the 352nd SOG from England and remain primed to effect rescues of downed airmen should they be needed. When Captain Scott O'Grady of the 555th FS was shot down in June 1995, it was a Marine TRAP team from USS Kearsarge which performed the recovery, for the simple reason that they were much closer to O'Grady's position than the AFSOC unit in the south.

Also at Brindisi are AC-130H gunships, which have been particularly active over Bosnia, especially during the early period of the conflict before SAMs were introduced to the region. These undertake armed reconnaissance missions by day or night, with the accent on night missions. Until fully night-capable tactical aircraft such as the F-15E and F/A-18D were deployed, the AC-130s were assigned to patrol Bosnia by night, while attack aircraft such as Jaguars and F-16s patrolled by day. On 10 March 1994 two AC-130Hs on patrol over Bosnia located targets which they prepared to attack, after French UN forces in Bihac were fired upon by Bosnian Serb tanks. Unfortunately, UN Command clearance came three hours later, by which time the Serbian forces had withdrawn. The AC-130Hs do not use their 2-kW searchlight, which is faired over, since it represents such a good IR source for enemy missiles, and the No. 1 gun, one of the 20-mm cannon, is removed to make way for an extra clear panel, behind which is an unidentified sensor. The AC-130H's 40- and 105-mm guns are cheaper than TOW or AGM-114 Hellfire missiles for taking out trucks or buildings, while the aircraft's ability to detect targets at extreme range, at night, makes it less vulnerable than attack helicopters.

With the growing number of limited conflicts and the growing importance of unconventional warfare, Air Force Special Operations Command no doubt has many more battles to fight. Its newly-delivered equipment and stream-lined organisation provide a solid base from which to build on the impressive combat record compiled since the Command gained its independence in 1990.

Randy Jolly

The AC-130H Spectre remains a vital element in AFSOC's force structure, despite its vulnerability over the modern battlefield. The SOFI improvement programme is increasing survivability and lethality, and will bring the AC-130H up to a level similar to that of the new-build AC-130Us now entering service with the 4th SOS.

US Air Force Special Operations Command

US Air Force special operations assets were previously concentrated within 23rd Air Force, but on 22 May 1990 the new Air Force Special Operations Command (AFSOC) stood up to take control of the force. This reflected the growing importance of special operations and unconventional warfare within the overall military structure, and allowed greater inter-service co-operation through the creation of a new specified command, Special Operations Command (SOCOM), to control the SOF forces of all four services.

AFSOC provides the Air Force component of SOCOM, and has under its control one wing based in the US and two groups overseas in Europe and the Far East. In addition it can draw on Air Mobility Command assets for greater airlift capability if required, and, indeed, some AMC C-130s, C-141s and C-5s can be configured to carry some of the equipment and undertake some of the 17th SOS AFSOC taskings.

In addition to the main flying units, AFSOC has a number of direct-reporting

units at its headquarters at Hurlburt Field. The 720th Special Tactics Group is comprised of special operations combat control teams (CCTs) and pararescue jumpers (PJs) who are literally the first in and the last ones out of practically any SOF or conventional conflict. Their mission includes surveying, establishing and controlling assault zones and forward area rearming and refuelling points, or FARRPS. The unit also calls in air strikes for the AC-130 Spectre gunships, artillery and naval gunfire.

The PJs provide emergency trauma care for wounded and injured personnel. Other important missions include mass casualty triage, establishing casualty collection and transfer points, conducting search and rescue operations and air-sea-land transport of survivors and patients. Special tactics teams routinely provide assistance in civic action and humanitarian affairs as well. The group has several units strategically located in the United States, Europe and the Pacific.

AFSOC also parents the 18th Flight Test Squadron. This was established in October 1983 as the Special Missions Operational Test and Evaluation Center (SMOTEC), and is the Air Force's only specifically tasked test and evaluation agency in support of enhancing Air Force special operations forces worldwide. The Flight Test Squadron's mission is to test and evaluate special operations aircraft. The unit was originally based at Edwards Air Force Base, California but was moved to Hurlburt Field, Florida in April 1994. It is arranged in a similar fashion to a flying squadron in that there is a commander and seven flights or divisions comprised of fixed-wing, rotary-wing, Combat Talon II, U-model gunship, electronic combat, operational analysis and operations support.

Also located at Hurlburt Field is the US Air Force Special Operations School. This is organised to educate selected US and allied personnel in the geopolitical, psychological, sociological and military considerations of AFSOC and US Special Operations

Command (USSOCOM). It also is the schoolhouse for US personnel involved in unconventional warfare and special operations missions. The school emphasises the human element. It is convinced that sensitivity to cultural differences is crucial since all special operations influence the behaviour of international allies and adversaries.

The USAF Special Air Warfare School was activated at Hurlburt Field in April 1967 and became a reporting unit to AFSOC in May 1990. During the early years the school's main emphasis was the preparation of Air Force personnel for duty in Southeast Asia. Since then, the school's curriculum has grown from a single course with 300 graduates a year to 15 different courses presented 72 times a year and graduating over 3,000 resident students. Courses offered include Third World orientation courses, joint operations courses, crisis response and international terrorism, psychological operations, and numerous joint operations courses.

16th Special Operations Wing/16th Operations Group

Hurlburt Field, Florida

The 16th Special Operations Wing, Hurlburt Field, Florida, is one of three Air Force special operations flying organisations. Its mission is to organise, train and equip Air Force special forces for worldwide deployment. The 16th SOW is the largest Air Force establishment under the Air Force Special Operations Command, which is the Air Force special operations component of the US Special Operations Command.

The 16th SOW manages a fleet of more than 90 aircraft with a military and civilian work force of nearly 7,000 people. There are seven flying squadrons. The wing was formed on 1 October 1993 at Hurlburt Field by renumbering the 1st SOW, which until then had been the primary Special Forces unit. Despite its illustrious history, the latter unit was inactivated to avoid the Air Force

having two active-duty flying wings with the same numerical designation, as the 1st Fighter Wing at Langley was considered to have had a superior pedigree. The wing has seven aircraft-operating squadrons (the 4th, 8th, 9th, 15th, 16th, 20th and 55th Special Operations Squadrons) with a mix of aircraft types, including the C-130E, AC-130H/U, MC-130E/H, HC-130N/P, NCH-53A, MH-53J and the MH-60G. All the assets of the 16th SOW are stationed at Hurlburt Field, apart from the HC-130s of the 9th SOS which are located at nearby Eglin AFB.

The wing parents the 16th Special Operations Support Squadron, which in turn controls a Special Operations Communications Flight (SOCF). The flight is comprised of special operations radio operators, communications maintenance

and communications planners. On the scene command and control is an integral part of special operations activities. To meet the demands of SOF operations, the communications flight's mission is to provide a fast-reaction, rapidly-deployable force capable of establishing and providing secure and non-secure voice and data command and control. The unit provides this via UHF satellite, HF, UHF, VHF communications systems in the field, mobile, shipboard and airborne configurations. The unit provides 24-hour reach-back capability between deployed SOF and Hurlburt Field commanders.

Aircraft type: C-130E (squadron assignment unknown)
Example aircraft: 62-1855, 63-7898

The 16th SOW is AFSOC's largest and most capable unit. Though based stateside, its aircraft are deployed worldwide.

4th Special Operations Squadron 'Puff', 'Spookies', or 'Ghostriders'

The 4th SOS was activated at Hurlburt Field on 5 May 1995 to relieve pressure on the 16th SOS, which until then was the only gunship squadron in the active-duty force. It is equipped with the Rockwell-modified AC-130U Spectre gunship.

The 4th was a pioneer in the gunship business. As the 4th Air Commando Squadron, it was activated at Forbes Field, Kansas in July 1965 with 11 C-47 and four FC-47 (later AC-47) aircraft. The 4th won distinction in the Vietnam War operating Douglas AC-47 'Spookies', which were the first fixed-wing gunships operated by the United States. It flew its last combat mission from Phan Rang on 30 November 1969 and was inactivated on 15 December 1969. Its AC-47s went to the South Vietnamese and Royal Laotian air forces.

As of July 1995, the 4th SOS was operational with nine of its intended 13

AC-130U gunships (representing the total USAF buy of these aircraft, including one attrition replacement for an AC-130H lost in the Gulf War). One of the nine (87-0128) aircraft remains attached to the AC-130U test force at Edwards AFB. The 19 crews of the 4th, most of them new to the gunship mission, were expected to be in training for the remainder of 1995 before the unit would be able to deploy or start operational missions and thus offer relief to the AC-130H-equipped 16th SOS.

Aircraft type: AC-130U
Example aircraft: 87-0128, 89-0510, 89-0511, 89-0511, 90-0163, 90-0164, 90-0165, 90-0166, 90-0167

The 'Ghostriders' are the proud possessors of AFSOC's 'U-boats' – the modernised AC-130U gunship.

8th Special Operations Squadron 'Black Birds'

Since its inception in 1917, the 8th SOS has flown more than 17 different type of aircraft and is currently the second largest continuous active-duty unit in the USAF.

In 1970, the Combat Talon was used in the assault on the North Vietnamese Son Tay POW camp. During the raid Combat Talon crews provided airborne jammer and command post duties, vectoring information to other aircraft involved in the mission. In April 1980, members of the 8th SOS were part of the mission to the Iranian desert in support of the American hostage rescue attempt. During that mission, five members of the squadron, flying a borrowed EC-130E configured as a tanker, were killed. The squadron gained its motto 'With the Guts to Try' from a package sent to the survivors by a British sister unit.

The 8th SOS participated in Operation Urgent Fury, the United States' October 1983 combined arms invasion of Grenada. Following the long flight from the US, the five 8th SOS Talons dropped Army Rangers, amid intense ground fire, precisely on time and on target. In December 1989, members of the 8th SOS were mobilised as part of the joint task force for Operation Just

Cause. Three of the squadron's MC-130Es were committed. A damaged MC-130E made the Hercules' first three-engined, short-field NVG take-off! At the end of the conflict in Panama, it was an 8th SOS MC-130E which was used to extradite General Noriega to the United States.

The MC-130E crews of the 8th SOS provided key aerial refuelling support for the SOF helicopters during Desert Storm, in addition to their much-publicised 'bombing' campaign featuring the 15,000-lb BLU-82 bombs. Members of the 8th SOS remain busy supporting ongoing activities in Operations Provide Promise and Deny Flight. It is believed that the re-equipment of the 1st SOS and 7th SOS with the MC-130H may eventually allow the concentration of the MC-130E fleet with the 8th SOS, although some aircraft may be passed on to the 919th SOS when it gives up its AC-130As.

Aircraft type: MC-130E-C
Example aircraft: 64-0523 (Mod 90), 64-0551, 64-0555, 64-0559 (Mod 90), 64-0561, 64-0562 (Mod 90), 64-0566 (Mod 90), 64-0567 (Mod 90), 64-0568

Right: An 8th SOS MC-130E lifts out of Nellis AFB, Nevada. The 'Black Birds' were responsible for delivering the huge BLU-82 bombs that devastated Iraqi positions during Operation Desert Storm. Today, the unit's MC-130E-Cs are undergoing the 'Mod 90' update.

Below: Along with the 'Mod 90' programme comes a new look for the 8th's aircraft. This two-tone grey scheme is slowly being adopted by AFSOC's C-130 variants

9th Special Operations Squadron 'Night Wings'

The 9th SOS was activated in April 1944 and originally designated as the 39th Bombardment Squadron. The squadron began operations at Grand Island Army Airfield, Nebraska and ended up at Tinian Island in the Marianas near Guam in December 1944. From Tinian the 39th BS conducted bombing raids on Tokyo and participated in dropping mines in the waters off the coasts of Japan and Korea. After World War II, in January 1946, the squadron moved to Clark Field, Philippines. In June 1947 the squadron moved to Kadena Air Base, Okinawa, and eventually inactivated in November 1948.

Reactivated again as a bomber unit on 2 January 1951, at Walker Air Force Base, New Mexico, the squadron began training with Boeing B-29s and later Convair B-36s. In December 1957, the first Boeing B-52 arrived and crews stood alert from

November 1958 through May 1959. In September 1963 the 9th BS was again inactivated.

Reactivated as the 9th Air Commando Squadron in January 1967, the squadron became the 9th Special Operations Squadron and flew O-2Bs and C-47s out of various locations in Vietnam including Nha Trang, Pleiku, DaNang and Bien Hoa. Psychological operations were the primary missions flown during the squadron's time in Vietnam.

The 9th SOS was inactivated in February 1972 and remained dormant until March 1988. Today the squadron is alive and well as an offspring of the 55th Aerospace Rescue and Recovery Squadron. When the 55th ARRS was redesignated the 55th Special Operations Squadron in March 1988, its HC-130 aircraft became the aircraft of the reactivated 9th SOS. The

9th SOS was a key player in Operation Just Cause, two of its HC-130s refuelling other 16th Special Operations Wing aircraft and allowing them to fly more than 400 missions for over 1,200 flying hours.

In August 1990, the 9th SOS deployed to Saudi Arabia in support of Desert Shield and prepared for the conflict by developing tactics and special mission rehearsals. During the war the squadron provided several refuelling and psychological leaflet drop missions. The 9th SOS continued on in support of the South-East Asia Cease-Fire Campaign after hostilities ended and did not depart the region until February 1993.

Today, the 9th Special Operations Squadron, based at Eglin Air Force Base, Florida, flies the HC-130N/P Combat Shadow aircraft. The 'HC' model is specially modified to support SOF helicopters in penetrating far behind enemy lines on clandestine missions at night and in poor weather. Since practically all of their missions are at night, the aircraft are compatible with night-vision goggles and allow the crews to perform their refuelling activities 'lights out'.

With their night-time capabilities and long range, the 9th SOS crews have become the squadron of choice for any large refuelling requirements and routinely refuel large formations of SOF helicopters. The 9th SOS has the world's only HC-130 Universal Aerial Refuelling Receptacle/Slipway Installment (USARRSI)-modified aircraft. This modification allows the HC-130 to refuel from KC-135 or KC-10 aircraft, providing almost unlimited range not only for the Combat Shadows but for the helicopters that they, in turn, refuel and support inbound and outbound from the target. The 9th SOS maintains a continuous overseas deployment supporting Operation Provide Comfort in Turkey.

Aircraft type: HC-130N
Example aircraft: 69-5819, 69-5828, 69-5831, 69-5832

Aircraft type: HC-130P
Example aircraft: 64-14854, 65-0991, 65-0993, 65-0994, 66-0213, 66-0217, 66-0220, 66-0225

15th Special Operations Squadron

The 15th Special Operations Squadron operates the MC-130H Combat Talon II. The Talon II is a derivative of the C-130H and is specially modified to support the special operations mission.

The 15th SOS started out as the 18th Observation Squadron in February 1942. In November of that year it was redesignated as the 15th Antisubmarine Wing based at Miami, Florida. The squadron was equipped with B-24s and participated in the extremely long-range over-water patrols in search of German U-Boats. The 15th Bomb Squadron (Heavy) was activated in June 1944 at Dalhart Army Airfield, Texas. It was equipped with specially modified B-29s, stripped of armament except for tail guns and fitted with the AN/APQ-7 'Eagle' radar which gave crews the ability to bomb through the clouds or in zero visibility. The 15th distinguished itself over Japan by bombing oil-producing facilities and, shortly thereafter, resupplying POW camps.

The 15th Air Commando Squadron was activated in March 1968, at Nha Trang Air Base, Vietnam, and assigned to the 14th Air Commando Wing. The unit had been known as Detachment 1, 314th Tactical Airlift Wing since 1966. Equipped with four WC-130Es, the unit conducted tactical airlift operations in support of selected American and Vietnam counter-insurgency forces in Southeast Asia under the Combat Spear programme. The unit performed numerous

Fulton recoveries and other dangerous missions. During one 90-day period unit personnel received two Silver Stars, 40 Distinguished Flying Crosses, 121 Air Medals, and eight Bronze Stars. The 15th Air Commando Squadron was renamed the 15th Special Operations Squadron in August 1968 and inactivated in October 1970.

Upgrades and expansion to the Combat Talon fleet began in 1981 and, in 1988, flight testing began on the MC-130H Combat Talon II. The first MC-130H arrived at Hurlburt Field, Florida in 1991 and was initially assigned to the 8th SOS. The 15th SOS reactivated in October 1992, allowing the 8th SOS to continue with the Talon I, and the initial cadre of squadron members continued the flight tests on the new Talon II aircraft while deploying and conducting formal school training. The unit has participated in over 24 readiness exercises and eight real world deployments, and successfully completed the first around-the-world deployment of a single Combat Talon II. The squadron is today the largest operator of the Talon II and acts as the *de facto* evaluation and training unit for the type.

Aircraft type: MC-130H
Example aircraft: 83-1212, 84-0475, 85-0011, 85-0012, 87-0024, 88-0195, 88-0264, 88-1803, 89-0280, 89-0281, 89-0282, 89-0283, 90-0161, 90-0162

Above: The 9th SOS was once a B-52 unit, but owes its present day incarnation to the 55th ARRS, whose HC-130s the 'Night Wings' inherited in 1988.

Below: 'SOFI' HC-130Ns, like this 9th SOS example, gain an undernose FLIR, while (in some cases) loosing the defunct Cook tracker dome.

16th Special Operations Squadron 'Ghost Riders'

The 16th Special Operations Squadron was for many years the only active-duty gunship squadron. In the 1980s and early 1990s, this suited USAF needs since gunships are deemed highly vulnerable to shoulder-mounted air defence missiles and other modern weapons. USAF plans to reduce and eventually retire its gunship fleet were confounded, however, by a post-Cold War world in which increased deployments to support operations in Somalia, Haiti, Bosnia and elsewhere made the AC-130 Spectre suddenly a much-needed item in inventory. Abruptly finding gunship crews among its 'most deployed' personnel (exceeding the intended maximum of 155 days deployed per year), the USAF was forced to scrap a plan to retire Reserve AC-130As, shift its AC-130Hs to the Reserve, and equip the 16th with AC-130Us. The decision was made to add the newer U models to the fleet by activating a new squadron (4th SOS).

Members of the 16th SOS therefore will continue to fly the AC-130H Spectre for close air support, armed reconnaissance, interdiction, night search and rescue and airborne command and control missions for SOF and conventional forces. The AC-130H is a highly modified C-130E configured with side-firing weapons including two 20-mm Vulcan cannons, a 40-mm Bofors canon, and one Army 105-mm Howitzer. On 30 October 1968, the 16th SOS Spectre became operational at Ubon Royal Thai Air Force Base as part of the 8th Tactical Fighter Wing. Equipped first with the AC-130A and later with more advanced AC-130E/H, Spectre crews continued in the tradition set forth by the AC-47 Spooky, AC-119G Shadow, and the AC-119K Stinger. In July 1974, Spectre crews moved as part of the 388th Fighter Wing to Korat Royal Thai Air Force Base, where that unit concluded its involvement in Southeast Asia. Having participated in practically every major campaign, the Spectre's final contribution in Vietnam was in supporting the evacuations of Saigon and Phnom Penh. It figured prominently also in the rescue of the captured SS Mayaguez. The Spectre's

distinguished record in Southeast Asia was not without cost, as 52 aircrew members were killed in action.

In December 1975, the gunships were moved to their present home as part of the 1st Special Operations Wing. Modified for inflight refuelling, the aircraft demonstrated their almost unlimited ranged with a 29.7-hour endurance flight from Hurlburt to Anderson AFB, Guam. The Spectres were back in action over Grenada in October 1983. Gunship crews provided surveillance and intelligence reports, enabling troop-carrying aircraft to be reconfigured for airdrop operations. Once US troops were on the ground, the gunships remained on station overhead where they provided extremely accurate cannon fire that knocked out numerous AAA emplacements, demolished several enemy armoured personnel carriers and shattered enemy troop concentrations.

Six years later the Spectre crews had another opportunity to demonstrate their prowess during Operation Just Cause in Panama, seven squadron aircraft being committed. The destruction of the Panamanian Defence Force headquarters – while attacking US troops were only a few metres away – exemplified the high level of training and skill of the gunship crews in providing close air support. Crews from the 16th SOS received both the MacKay Trophy and the Military Airlift Command Air Crew of the Year Award for 1989 for their efforts in Just Cause. In September 1990, the 16th SOS arrived in Saudi Arabia in preparation for Operation Desert Storm. Although 50 combat missions were flown, the loss of 'SPIRIT 03' and its crew of 14 was the costliest aircraft loss of the war and a reminder of the risks gunship crews face.

Since the Gulf War, 16th SOS crews have been deployed almost non-stop. During 1993 and 1994, they deployed to Africa in support of Continue Hope, United Nations relief operations to Somalia. During this deployment, 'JOCKEY 14' was lost when an in-bore detonation of the 105-mm gun occurred while airborne. Eight of the 14 crew died as the aircraft ditched in the ocean near Mogadishu.

The AC-130H gunships from the 16th SOS today fly in Operation Deny Flight and have flown numerous direct support missions over Bosnia-Herzegovina.

Below: The camouflage scheme worn by the Spectres of the 16th SOS today is a far cry from that of their Vietnam days.

Aircraft type: AC-130H
Example aircraft: 69-6568, 69-6569, 69-6570, 69-6572, 69-6573, 69-6574, 69-6575, 69-6577

Above: Throughout 1995 16th SOS AC-130Hs have been deployed to Brindisi as part of the ongoing Operation Deny Flight.

20th Special Operations Squadron 'Green Hornets'

The 20th SOS, based at Hurlburt Field, flies the MH-53H/J Pave Low III Enhanced as its main mission equipment. Two NCH-53As are also assigned for trials work. The 'Green Hornets' conduct day or night low-level missions to insert and extract special forces teams. These 'customers' are most often Army Rangers, Navy SEALs or combat control teams (CCTs) and pararescue (PJ) teams. Because of the Pave Low's great range and its ability to fly extremely low night-time operations, it has become the workhorse of special ops.

The 20th SOS was formed as the 20th Observation Squadron (Light) at Savannah Air Base, Georgia in March 1942, flying A-20, B-25, DB-7, L-1, L-5, P-40 and P-43 aircraft. The squadron was redesignated the 20th Reconnaissance Squadron in April 1943 at Vichy, Missouri. The 20th first saw combat in January 1944, flying out of Fushkara, India in support of the 1st Air Commando Group. Back in the US at the end of World War II, the squadron was inactivated in November 1945.

Reactivated in 1956 as part of Tactical Air Command, the 20th Helicopter Squadron flew the H-21 helicopter. By the end of 1965, the unit was flying CH-3s in Vietnam, using them to run Pony Express clandestine drops and pick-ups in North Vietnam and Laos. In 1967 the CH-3s were joined by UH-1F/Ps and continued to perform unconventional warfare missions until 1972, when the unit was again inactivated.

In 1976, the 20th was reactivated at Hurlburt Field operating UH-1N and CH-3 gunships. In 1980 the first HH-53s arrived, replacing the ageing CH-3s and ushering in a new era of heavy-lift capability. Between 1983 and 1985 the UH-1Ns participated in Operation BAT (Bahamas, Antilles and Turks) as part of the South Florida Drug Enforcement Task Force. MH-53Hs arrived in 1986, and MH-53Js in 1988. After seeing limited action in Operation Just Cause with five aircraft, the 20th SOS made significant contributions to the war against Iraq. As has been reported, Pave Low crews were afforded the distinction of 'starting the war' when they led Army 'Task Force Normandy' AH-64s to attack Iraqi radar sites during the opening minutes of the air war. In addition to effecting the dramatic rescue of a downed US Navy pilot, the Pave Low crews also ferried special operations teams to and from their destinations far behind Iraqi lines.

The 20th SOS also has considerable involvement in Operations Southern Watch, Provide Comfort, Provide Promise, Provide Comfort II and Support Democracy.

Aircraft type: MH-53J
Example aircraft: 66-14431, 67-14993, 67-14994, 67-14995, 68-8284, 68-8286, 68-10356, 68-10357, 68-10358, 68-10360, 68-10363, 68-10364, 68-10923, 68-10924, 69-5785, 69-5789, 69-5790, 69-5791, 69-5793, 69-5794, 69-5795, 69-5796, 69-5797, 70-1629, 70-1630, 70-1631

Right: The 20th Special Operations Squadron, the 'Green Hornets', is the 16th SOW's MH-53J Pave Low III component, and is the largest helicopter-equipped unit in AFSOC.

55th Special Operations Squadron 'Night Hawks'

The 55th Special Operations Squadron, operating out of Hurlburt Field, is the largest H-60 squadron in the Air Force. The 55th SOS Pave Hawks are highly sophisticated and modified to meet special operations demands. They incorporate a FLIR integrated with a global positioning system and an inertial and Doppler navigation system, giving the Pave Hawk incredible accuracy and on-time delivery of special operations troops and supplies. Since the MH-60G can be refuelled in flight, its range is almost unlimited. Armed with either two 7.62-mm or two .50-calibre machine-guns (or a combination of both), the Pave Hawk can fight its way into and out of hostile territory and provide fire support for the special operations forces it delivers or picks up. Because of its small size, the Pave Hawk has demonstrated its ability for shipboard as well as land operations.

The 55th SOS was originally activated as the 55th Aerospace Rescue and Recovery Squadron in November 1952 at Thule Air Base, Greenland and remained there until inactivated in 1960. The squadron was reactivated in February 1970 and moved to McCoy Air Force Base, Florida. In June 1971, the 55th moved to its present location at Hurlburt Field.

In December 1982, the squadron received its first four UH-60 Black Hawk helicopters to replace its H-3s. Interface with special operations had already begun and numerous avionics and equipment modifications including aerial refuelling were adopted. The unit became the 55th SOS in March 1988.

The 55th SOS MH-60G Pave Hawks are the only truly rapidly deployable helicopters in the US Air Force inventory. Four fully equipped Pave Hawks can be loaded into a

C-5 and transported anywhere in the world. The squadron demonstrated its rapid deployment capability during the 1989 search in Ethiopia for Congressman Mickey Leland, the victim of a civil aircraft crash. The Pave Hawks were deployed in just 14 hours after the initial notification of their involvement.

Aircraft type: MH-60G
Example aircraft: 87-26006, 87-26007, 87-26008, 87-26009, 87-26010, 87-26011, 87-26012, 87-26013, 87-26014, 89-26204

Right and below: The story of the MH-60G is a convoluted one. Today's SF-configured MH-60Gs closely resemble HH-60G 'Rescue Hawks' but for their AAQ-16 FLIR and armament. HH-60Gs can carry 7.62-mm guns but only the MH-60G carries the 0.50-in machine-gun.

193rd Special Operations Group

Harrisburg IAP, Middletown, Pennsylvania

During the 1960s, four ANG squadrons were assigned a special operations mission, equipped with U-10s, HU-16s and C-119s. This tasking ended in 1975, leaving the 193rd SOG as the only ANG special operations unit. The 193rd SOG transferred to MAC (from TAC) in March 1983, transferring again to the newly-formed AFSOC in May 1990. This change in assignment was accompanied by adoption

of the Commando Solo codename for the mission, in place of the Volant Solo codename used previously. Technically, the 193rd is controlled in peacetime by the state (of Pennsylvania), but in time of war would be gained by AFSOC. Due to its unique nature, the unit is continually involved in missions tasked directly by AFSOC/SOCOM and is deployed on active duty for a large proportion of its time.

193rd Special Operations Squadron 'Quiet Professionals'

The 193rd SOS is the group's operational squadron and has flown the EC-130E since 1977. The unit was formed in September 1967 at Olmstead AFB (now Harrisburg IAP) as the 193rd Tactical Electronic Warfare Squadron with the C-121C. The specialist EC-121S performed Coronet Solo missions, and two were dispatched to Southeast Asia in the latter half of 1970 under the codename Commando Buzz. The 193rd was redesignated a Special Operations Squadron from April 1977 until December 1977, when it reverted to TEWS

status. The unit began conversion to the EC-130E in 1977. Designation was changed to the 193rd Electronic Combat Squadron in October 1980 but reverted to the 193rd SOS in mid-1984. The 193rd participated in operations in Grenada (1983), Panama (1989) and the Persian Gulf (1991). It operates EC-130E(RR) Rivet Rider and EC-130E(CL) Comfy Levi EW variants of the C-130. Comfy Levi is an Elint and military communications jamming/broadcasting configuration, while River Rider intrudes on or replaces an adversary's civilian TV and radio broadcasts. All aircraft have the Universal Aerial Refuelling Receptacle/Slipway Installment (USARRSI).

Aircraft type: EC-130E (CL)
Example aircraft: 63-7815, 63-7816, 63-7828, 63-9816

Aircraft type: EC-130E (RR)
Example aircraft: 63-7773, 63-7783 – former EC-130E(CL), 63-7869, 63-9817. All now to worldwide colour TV standard

Aircraft type: C-130E
Example aircraft: 63-9815 – possibly now converted to EC-130E(CL) standard

The once highly secret Comfy Levi/Senior Scout antenna fit has now been seen on several C-130s, not just on the EC-130E(CL)s serving with the 193rd SOS, and indeed can be carried by any Hercules modified to Senior Hunter standards to accept the palletised mission equipment. The weight of the installation necessitates using a C-130 with C-130H-type engines, and operationally the mission tends to require inflight refuelling. This aircraft belongs to the 463rd AW.

352nd Special Operations Group

RAF Mildenhall, UK

The 352nd SOG is headquartered at RAF Mildenhall, England. It conducts special operations and combat rescue throughout Europe, Africa and the Middle East.

The 352nd SOG activated on 1 December 1992 and replaced the 39th Special Operations Wing. All associated squadrons moved from Rhein Main Air Base, Germany and RAF Woodbridge, UK to consolidate AFSOC Forces at RAF Alconbury. As part of the United States Air Forces in Europe (USAFE) force restructuring the group moved from RAF Alconbury to RAF Mildenhall on 17 February 1995.

The 352nd SOG traces its heritage back to the 2nd Air Commando Group which was activated on 22 April 1944 at Drew Field, Florida. Personnel from two fighter squadrons, the 1st and 2nd Fighter Squadrons (Commando), started training for their Air Commando operations. On 1 May 1944, three Liaison Squadrons (Commando) – the 127th, 155th and 156th – teamed up with the 317th Troop Carrier Squadron (Commando) to join the 2nd ACG.

In October 1944, the group began its move to India. Crews of the 317th TCS(C) flew their C-47s to India, arriving in late October. They began almost immediately dropping supplies to Wingate's Raiders who were fighting the Japanese in the Chindwin Valley of Burma. The unit later flew Chinese troops from Burma to China, transported men, food, ammunition and construction equipment to Burma, and dropped Gurka paratroops during the assault on Rangoon.

The pilots of the 127th, 155th, and 156th LS(C) began operations in January 1945, and performed reconnaissance, light transport and evacuation duties with their L-4s, L-5s and C-64s. The 1st and 2nd FS(C), flying the P-51, provided fighter support after arriving on the scene in February 1945.

The 127th, 155th and 156th LS(C) left the 2nd ACG for duty in the Pacific in May 1945, and the 317th TCS(C) ended its association with the group in September. By October the group was headed home, and in November 1945 it was inactivated at Camp Kilmer, New Jersey.

The group was activated at Alconbury in December 1992 to replace the 39th SOW. The unit brought together the special forces elements, including the HC-130s and MH-53s which were previously stationed at Woodbridge and the MC-130s from Rhein Main, Germany. However, the stay at Alconbury was comparatively short-lived as the unit began moving to Mildenhall during the winter of 1994/spring of 1995. The MC-130Hs of the 7th SOS were the first to relocate. The squadron had completed the move by mid-January 1995, with its aircraft parked along the old cross runway on the

south side of the base. Most of the 67th SOS HC-130N and P models transferred to Mildenhall during the first week of January. They were followed by the MH-53Js of the 21st SOS, which arrived by 17 February as the final stage in the relocation of Europe-based Air Force Special Operations Command forces. The move by the 21st SOS was far more protracted that the other two squadrons, as the MH-53Js did not complete their transfer for several weeks.

Today, the 352nd SOG is the air component command for special operations within the European Command. Under the operational control of Special Operations Command Europe (SOCEUR), the group plans and executes general war and contingency operations using specialised aircraft, tactics and air-refuelling to infiltrate,

exfiltrate and resupply special operations forces. The unit has six squadrons, three of them operating aircraft, and four different types of aircraft assigned at Mildenhall.

In the last five years the group or its squadrons have deployed to support Operations Desert Storm, Provide Hope, Silver Anvil, Provide Promise and Deny Flight.

The 352nd Operations Support Squadron provides the group with current flying data, up-to-date intelligence and the co-ordination to operate from multiple locations. The 352nd Maintenance Squadron provides all intermediate-level maintenance and support on MC-130H, MH-53J and HC-130N/P aircraft assigned to the three flying squadrons. The 321st Special Tactics Squadron provides a fast-reaction, rapidly deployable force capable of establishing and providing air traffic control and air/ground communications in hostile territory under covert, clandestine or low-visibility conditions. The combat control teams (CCTs) and pararescuemen (PJs) are usually the first ones in to set up a landing zone, reconnoitre the area in preparation for the arrival of other special forces and provide casualty treatment. The teams are capable of employing high-altitude, low-observability (HALO) and static line parachute drops, SCUBA, watercraft or any other method required by the mission.

The 352nd Special Operations Group also parents a Special Operations Communications Flight. For more details of these flights' duties, see the entry under 16th SOW.

7th Special Operations Squadron 'Aircommandos'

The 7th SOS operates five MC-130Hs. Their missions usually involve night or adverse-weather, long-range, low-level infiltration and exfiltration and resupply operations. With the unique capabilities of the Talon II, aircrews using their terrain-following radar, precision navigation equipment and electronic countermeasures support ground and maritime special operations forces in potentially hostile or denied territory. The squadron completed its move to Mildenhall from Alconbury by February 1995.

Aircraft type: MC-130H
Example aircraft: 84-0476, 86-1699, 87-0023, 88-0193, 88-0194

21st Special Operations Squadron 'Dust Devils'

Having previously flown HH-53Cs from RAF Woodbridge, the 21st SOS now flies five MH-53J Enhanced Pave Low IIIs on night and adverse-weather missions to support special operations land and amphibious forces, and maintains a 24-hour all-weather CSAR commitment. The 21st completed its move from Alconbury to Mildenhall on time on 17 February 1995, despite maintaining a detachment in Italy for Balkan operations.

Aircraft type: MH-53J Enhanced Pave Low III
Example aircraft: 68-10930, 69-5784, 70-1625, 70-1626, 76-1648

Above: The 7th SOS undertakes European MC-130H operations and have been active over Bosnia from their Mildenhall home.

Below: Nose art is common on the AC-130, but rarely seen on the MC-130. This 7th SOS MC-130H sports a suitably macabre decoration.

Above: The 21st SOS was once part of the 67 ARRS at RAF Woodbridge, before adopting an SF role.

Below: Along with aircraft from the 20th SOS, the 21st SOS currently maintains a detachment at Brindisi.

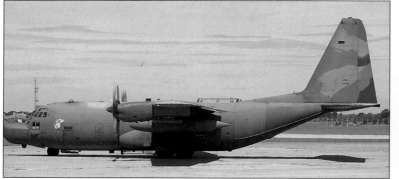

67th Special Operations Squadron 'Night Owls'

The 67th SOS operates five Lockheed HC-130N/P tanker aircraft which provide tanker support for the Pave Lows. The squadron also keeps a single C-130E Hercules for routine transport missions. Night-vision goggles and low-level tactics give the 67th SOS crews the ability to perform their refuelling and resupply missions at night or in adverse weather.

Aircraft type: HC-130N
Example aircraft: 69-5820, 69-5823, 69-5826

Aircraft type: HC-130P
Example aircraft: 66-0223

Aircraft type: C-130E
Example aircraft: 63-7814

Above right: When the 67th ARRS was disbanded, and its aircraft transferred to the special operations role, the HC-130 Combat Talons remained with the newly renamed 67th SOS, while HH-53 operations (now MH-53J) were passed to the newly-formed 21st Special Operations Squadron.

Right: This 'slick' C-130E is used as a 'bounce bird' for crew training by the 67th SOS. It was previously on charge with the 7th SOS, when that unit was based at RAF Alconbury, along with the rest of the 352nd SOG.

353rd Special Operations Group

Kadena AB, Okinawa

The unit was established as the 353rd SOW in 1990 as the air component for special operations in the western Pacific region, operating the MC-130E, MH-53J and HC130N/P. The 17th SOS had previously operated as an autonomous unit, and was joined within the new wing by the 38th ARRS. The wing was stationed at Clark AB with a squadron based at Kadena AB. The eruption of Mount Pinatubo in mid-1991 forced the USAF to vacate Clark and move to NAS Cubi Point initially, before relocating to Kadena. The wing was downgraded to group status in December 1992. The 31st SOS is stationed at Osan AB, South Korea.

The group's mission is to act as the focal point for all US Air Force special operations activities throughout the Pacific, but also maintains a vital CSAR commitment. The 353rd SOG has three flying squadrons, and maintenance, tactical communications and special tactics squadrons. The 320th Special Tactics Squadron pararescuemen and combat controllers provide for the establishment of drop and landing zones, provide air traffic control, combat medical care and evacuation, and C-SAR

The 353rd Special Operations Group also parents a Special Operations Communications Flight. For more details of these flights' duties, see the entry under 16th SOW.

1st Special Operations Squadron

The 1st Special Operations Squadron has been a long-term operator of the MC-130E Combat Talon I, but may be undergoing conversion to the MC-130H. Their mission is to deliver troops and equipment by airdrop or by landing at austere sites behind enemy lines at night and in adverse weather.

Aircraft type: MC-130E-Y
Example aircraft: 62-1843, 63-7785 (Mod 90), 64-0565 (Mod 90)

Aircraft type: MC-130E-S
Example aircraft: 64-0571, 64-0572

Right and below: This 1st SOS MC-130E was one of the first 'Mod 90' Combat Talons to be converted and was suitably adorned.

17th Special Operations Squadron 'Jackals'

The 17th SOS operates the HC-130N/P Combat Shadow and is responsible for air refuelling of special operations helicopters. The crews are night-vision goggle trained and the vast majority of their missions are night-time, low-altitude refuelling of special operations helicopters, although the squadron is fully capable of dropping or landing equipment and supplies.

Aircraft type: HC-130N
Example aircraft: 69-5822

Aircraft type: HC-130P
Example aircraft: 64-14858, 65-0992, 66-0215

Aircraft type: C-130E
Example aircraft: 63-7842

The 17th SOS provides tanker support for the other aircraft of the publicity-shy 353rd Special Operations Group. The group (formerly a wing) moved its headquarters to Kadena from Clark AFB, in the Philippines, after the US pull-out from those islands in 1991/92. Now the HC-130s and MC-130s of the 17th and 1st SOS share their Okinawa home with the F-15s, E-3s, KC-135s and HH-50s of PACAF's 18th Wing.

AFSOC units

31st Special Operations Squadron 'Black Knights'

The 31st Special Operations Squadron is the only AFSOC component on the Korean peninsula. Located at Osan Air Base, Korea, the squadron operates the MH-53J Enhanced Pave Low III for behind-the-lines infiltration and exfiltration missions under cover of darkness or bad weather. The squadron is also available for search and rescue, although the HH-60G-equipped 38th RS at the same base (which is not part of the special operations community) has primary responsibility for that role.

Aircraft type: MH-53J
Example aircraft: 68-10367, 68-10923, 68-10926, 68-10932, 73-1649

The 31st SOS is the sole 353rd SOG unit not based at Kadena.

427th Special Operations Squadron

Pope AFB, NC

The 427th SOS, previously known as Det. 6 of HQ AFSOC, operates the CASA C.212-200 and -300. These aircraft apparently support the US Army and joint-service special operations community at adjacent Fort Bragg and were deployed to the Persian Gulf during or immediately after Operation Desert Storm. The CASA transports, which do not appear on the official USAF inventory, are apparently also used as a low-cost method of providing jump opportunities to Army and Air Force personnel maintaining their parachute qualifications.

Aircraft type: CASA C.212-200
Example aircraft: 87-0158, 87-0159, 90-0168, 90-0169

The elusive Aviocars of the 427th Special Operations Squadron have a unique place in the AFSOC line-up. The Spanish-built aircraft provide an unobtrusive method of carrying out a small airdrop of personnel or material that the MC-130s, despite all their sophistication, can never match.

919th Special Operations Wing/919th Operations Group

Duke Field, Florida

The 919th was activated as a troop carrier group in February 1963 at the Memphis Municipal Airport as part of the Air Force Reserve. It became a tactical airlift group in July 1971, relocating to Duke Field. In October 1971 the 919th TAG received the first of six C-130As. The group used the rare RC-130A version, although these had been demodified to standard transport configuration. The unit converted to the C-130B in August 1973 and tactical training began in October 1973.

In 1974 it was announced that the unit would convert to the AC-130A gunship. Aircrew began conversion with the 1st SOW at Hurlburt, ready for the arrival in June 1975 of the first of 10 AC-130As. The 919th was redesignated as the 919th SOG in July 1975 (the only AFRes Special Operations Group), achieving combat readiness in July 1976. Dedication and training earned the reservists a second

Outstanding Unit Award for meritorious service between July 1975 and January 1977. The unit had claimed four of these by 1989.

Two of the 919th's AC-130As were at Howard AFB at the start of Operation Just Cause, and were positioned long before the active-duty gunships could fly down from Hurlburt. The 919th was called to active duty for Operation Desert Storm, and on 30 June five AC-130As flew out to the Middle East, where they amassed 125 combat hours. Three aircraft contributed to the carnage on the Basra Highway during the 'Mother of all Retreats'. The 919th SOG was elevated to wing status in June 1992. In mid-September 1994, the unit participated in Operation Uphold Democracy in Haiti, becoming the first tactical aircraft over Port-au-Prince, and remained in the area until December 1994 as peacekeepers. Four AC-130As were retired during this period. This Reserve component is undergoing considerable

change, having stood up as the 5th Special Operations Squadron with HC-130P/Ns and facing the long-term prospect of changes for its 711th SOS.

The peacetime mission of the 919th SOW is to train Air Force reservists on the AC-130A, HC-130N/P and MC-130E/H in aircraft operations, maintenance and support functions. Wartime missions would include air operations in support of conventional and unconventional warfare, the provision of offensive and defensive firepower in support of friendly forces, close air support of troops in contact, armed reconnaissance and the interdiction of enemy lines of communication. The wing reports to the Air Force Reserve's Tenth Air Force at Bergstrom, but when mobilised is gained by Air Force Special Operations Command at Hurlburt. The 919th has more than 1,400 reservists and 300 full-time civilian employees, with six AC-130As, two C-130Es and a planned establishment of five HC-130s. The reservists each attend

one weekend training assembly per month and perform 15-day active-duty training periods. The 919th SOW controls three groups, the 919th Logistics Group with the 919th Maintenance and Logistics Support Squadrons, the 919th Support Group with the 919th Civil Engineer, Security Police and Mission Support Squadron, and the 919th Operations Group with the flying squadrons. The 919th Medical Squadron reports directly to the wing.

5th Special Operations Squadron 'Shadows'

The 5th SOS at Duke Field, Florida activated on 7 January 1995 and operates the HC-130N/P Combat Shadow, although some sources suggest that its crews are also training in the Combat Talon role. This has not been officially confirmed or explained. The squadron's mission is to conduct clandestine intrusion of enemy territory to support, resupply, insert and recover special forces, and to provide aerial refuelling for special operations helicopters. Most personnel came from the 711th SOS, when that unit lost four of its ageing AC-130A gunships, and the unit initially borrowed aircraft from the 9th SOS. The squadron participated in a three-ship

formation which refuelled eight MH-53s during February 1995, and began airdrop training in April. On 11 April a 5th SOS Hercules airlifted six 160th SOAR AH-6s from Fort Campbell to Hill AFB, and was said to have subsequently flown air-to-air refuelling missions in support of the Army special forces helicopters, although it was not previously acknowledged that the AH-6 had an inflight-refuelling capability. The squadron will probably conduct a great deal of tasking in support of the 160th, particularly supporting the regiment's permanently probe-equipped MH-60s and MH-47s. By the end of June the squadron had accumulated over 100 flying hours in 30 sorties, and had off-loaded 80,000 lb (36290 kg) of fuel. Also during April the squadron received the first four of its aircraft from various active-duty and AFRes units, having previously borrowed aircraft from the 9th SOS. A fifth aircraft arrived later in 1995. Rumours persist that the squadron will receive a handful of MC-130Es.

Aircraft type: HC-130N
Example aircraft: 65-0971 (ex HC-130H), 69-5825, 69-5827

Aircraft type: HC-130P
Example aircraft: 66-0216, 66-0219

711th Special Operations Squadron 'Spectres'

This squadron began life in July 1971 as the 711th TAS and reported to the 59th TAW until December 1974, when it was reassigned to the 919th TAG. It initially operated the Lockheed C-130A Hercules and upgraded to the C-130B in July 1973, becoming combat ready with the new type within seven months. The 711th converted to the Lockheed AC-130A Spectre early in 1975 as the 711th SOS. It currently operates as part of the 919th SOW.

The 711th has now operated the same aircraft type and model longer than any other flying squadron in the USAF. Among its aircraft are 53-3129, the first operational Hercules, known as *First Lady*. Long-standing plans for the squadron to receive the AC-130H models belonging to the 16th SOS have been cancelled and the unit is now due to lose the gunship mission altogether. Four AC-130As were retired in autumn 1994, but six remained in use as the unit transitioned to the Combat Talon mission using MC-130Es borrowed from the 8th SOS, and to the Combat Shadow mission using aircraft from the 5th SOS. The unit may eventually pick up MC-130Es from other units, or its crews may continue to rely on borrowed aircraft. Alternatively the squadron may disappear altogether, and its crews may join the 5th SOS. *First Lady* was scheduled to be the last AC-130A to be

retired, with a ceremony planned for 10 September 1995.

Aircraft type: AC-130A
Example aircraft: 53-3129, 54-1623, 54-1628, 54-1630, 55-0014, 56-0469

Aircraft type: C-130E
Example aircraft: 65-0969, 65-0972

Aircraft type: C-130H
Example aircraft: 64-4859

Aircraft type: C-130A (possibly retired)
Example aircraft: 56-0522, 57-0469

Exterminator is one of the most famous of the 711th SOS's AC-130As, having fought in every major action since Vietnam. The AC-130As have not received a SOFI-type upgrade and are now being retired.

Finally retired in September 1995, AC-130A 53-3129 First Lady was the first production Hercules. It was modified to AC-130 standard in 1970 and served with the 711th SOS for nearly 20 years.

Air Education and Training Command

58th Special Operations Wing/58th Operations Group

Kirtland AFB, New Mexico

The history of the 58th Wing and of the training establishment at Kirtland followed separate routes until they converged on 1 January 1993. On that date Air Education and Training Command took over aircraft type training for the MH-53, MC-130 and HC-130, a task that had formerly been performed under Air Mobility Command (AMC).

The training establishment began as the 1550th Aircrew Combat Training and Test Wing which moved to Kirtland from Hill AFB, Utah in 1976. Rooted at Kirtland is a long tradition of nurturing aircrews for what was known then as the Aerospace Rescue and Recovery Service, which was part of Military Airlift Command (changed to Air Mobility Command on 1 June 1993). In May 1984, this training wing became the 1550th CCTW (Combat Crew Training Wing). On 1 October 1991 the 1550th merged with the 1606th Air Base Wing to create the 542nd CCTW, which was briefly assigned to AFSOC. On 1 January 1993, the wing split again and the 542nd went to AETC which took over all type training from the combat commands. As part of an effort to preserve the traditions of certain combat establishments, on 8 April 1994 the 542nd

CCTW was renamed 58th SOW, even though all previous 58th history involves fighters.

Today's 58th carries the traditions of the 58th Pursuit Group, constituted on 20 November 1940 and equipped initially with Curtiss P-36s. The group flew P-39s and P-40s before going to the Pacific in Republic P-47 Thunderbolts. The group fought at New Guinea, Mindoro and Okinawa and was one of the final users of the Thunderbolt before being inactivated on 27 January 1946. The 58th Fighter-Bomber Wing was activated at Taegu AB, Korea on 10 July 1952, when it merged two existing fighter groups equipped with the Republic F-84E/G Thunderjet.

This combat wing operated at several bases in the Far East until inactivated in July 1958. The wing was redesignated 58th Tactical Fighter Training Wing and activated at Luke AFB in October 1969, replacing the 4510th CCTW training F-100 Super Sabre and later F-4 Phantom aircrews. A squadron of A-7Ds was assigned between 1969 and 1971, and the F-5 training unit was located at Williams AFB from 1969 until transferred to 405th TTW in August 1979. The unit began converting to the F-15A in 1974 and

changed designation to 58th Tactical Training Wing in April 1977. F-15 training was transferred to the 405th TTW in August 1979, with the 58th TTW concentrating on the F-4C until 1983, when the F-16A commenced delivery. The wing began receiving the F-16C in 1984, and was redesignated 58th Fighter Wing in October 1991 with F-15Es transferred from the 405th TTW to consolidate training duties at Luke AFB. The wing was replaced at Luke AFB by the 56th FW, and its identity went to Kirtland.

Kirtland AFB is in Albuquerque (a portion of its territory is used by the Albuquerque Airport). The base was included on a tentative BRAC (Defense Base Realignment and Closure Commission) report in early 1995 which seemed to number its days and created very tentative plans for the 58th Special Operations Wing to move to the already seriously overcrowded Holloman AFB a few miles to the south at Alamogordo. BRAC has since spared Kirtland and no current plans exist for the 58th to move. The wing's 512th, 550th and 551st Special Operations Squadrons fly a mix of HC-130N/Ps, MC-130Hs, MH-53Js and MH/HH-60Gs to train rescue aircrew, as

well as those of Air Force Special Operations Command (AFSOC). The wing's mission is to train aircrew members in advanced helicopter, MC/HC-130, pararescue and combat control techniques, but also provides personnel and airlift capacity in response to crises around the world, and assists civilian authorities in regional rescues. The training mission includes classroom instruction, simulator training and flying exercises. These flying missions include conversion, transition, instrument flying, inflight refuelling, and personnel and equipment airdrops. The wing also provides tactical training (including use of NVGs) and rescue hoist training.

In addition to training aircrew, the 58th SOW is also responsible for training crewmen for their pararescue (PJ) and combat control team (CTT) qualifications. In addition to rigorous physical training, all students receive instruction in mountaineering, parachuting, survival, escape and evasion, signalling, weapons, scuba diving and overland navigation. PJ students learn the practical applications of advanced medical procedures, allowing them to treat trauma and perform surgical procedures until more extensive medical attention is available. CTT students learn air traffic control, advanced command, control and communications, drop/landing/recovery zone preparation, fire control and demolition.

512th Special Operations Squadron

Known until 8 April 1993 as the 1550th TTS (Tactical Training Squadron), the 512th SOS operates the UH-1N Huey, HH-60G Hawk, and MH-60G Pave Hawk helicopters.

Aircraft type: UH-1N
Example aircraft: 69-6603, 69-6610,

69-6612, 69-6627, 69-6649, 69-6650, 69-6665

Aircraft type: MH-60G
Example aircraft: 81-23644, 81-23646, 82-23671, 82-23680, 88-23689, 82-23708, 82-23718

550th Special Operations Squadron 'Wolfpack'

Designated the 1550th FTW (Flying Training Squadron) prior to 8 April 1993, the 550th SOS is the only squadron in the 58th SOW with a familiar nickname. The squadron operates the MC-130H and HC-130N/P.

Aircraft type: HC-130N
Example aircraft: 69-5821

Aircraft type: HC-130P
Example aircraft: 65-0975, 66-0212,

Aircraft type: MC-130H
Example aircraft: 87-0125, 87-0126, 88-0191, 88-0192

551st Special Operations Squadron

Known as the 1551st until 8 April 1993, the 551st is the USAF training squadron for the MH-53J Enhanced Pave Low III helicopter. The squadron operates stripped-down TH-53As (ex-Marine Corps CH-53As) as well as its fully missionised MH-53J Enhanced Pave Low IIIs.

Aircraft type: TH-53A
Example aircraft: 66-14468, 66-14470, 66-14471, 66-14472

Aircraft type: MH-53J
Example aircraft: 66-14428, 66-14429, 66-14432, 66-14433

Above: The 'Wolfpack' operates a mix of C-130 variants, including HC-130N/P tankers.

Below: Pave Low conversion training is the task of the 551st SOS with a fleet of TH-53s and NCH-53s.

The MH-60s flown by the 512th SOS are not full-standard Pave Hawks, but partner elderly UH-1s in the helicopter training role

F-16 Operators: Part 1

US Air Force

Since the first aircraft was delivered in 1979, the USAF has received about 2,200 F-16s. By the start of 1995 some 1,640 were still in the inventory, with most of the F-16A/Bs from early block numbers having been retired. Of the total force, 777 were assigned to active-duty units, 742 to the Air National Guard and 121 to the Air Force Reserve. During the early 1990s the USAF underwent a massive organisational upheaval coincident with a considerable draw-down in force size and a move to modernise the reservist forces, and recent changes to the F-16 operators reflect this. Details of deactivated units are also presented here for completeness.   David Donald

Air Combat Command (Tactical Air Command)

9th Air Force

20th Fighter Wing
(formerly 363rd Fighter Wing)

Shaw AFB, South Carolina 'SW'

17th Tactical Fighter Squadron
(white - not current)
19th Tactical Fighter Squadron
(gold – not current)
33rd Tactical Fighter Squadron
(blue – not current)
77th Fighter Squadron (red)
78th Fighter Squadron (yellow/red)
79th Fighter Squadron
(yellow/black)

From its establishment in the independent Air Force in 1947 the 363rd was principally a reconnaissance unit, and has been at Shaw since 1951. In April 1982 it began to convert from 'JO' coded RF-4Cs to the F-16A/B Block 15, adopting a dual-role mission with the accent on attack. Two squadrons were equipped with the Fighting Falcon: the 17th TFS 'Owls' (later 'Hooters') and 19th TFS 'Gamecocks'. The first-generation jets were soon replaced by F-16C/D Block 25s. One squadron of RF-4Cs was retained until September 1989 until becoming an F-16C operator (33rd TFS 'Falcons'). These units carried their nicknames within the coloured fin-bands, while the tailcode changed to 'SW'.

In 1990 the wing established the 363rd Tactical Fighter Wing (Provisional) at Al Dafra AB, Sharjah, for Operation Desert Storm, and dispatched the 17th TFS and 33rd TFS to be the core of the wartime unit. The 363rd squadrons were in the thick of the fighting, the 17th TFS also operating a handful of aircraft transferred from the 50th TFW with the non-standard 'DS' tailcode. A 33rd TFS aircraft was painted for a week in a sand camouflage, but this was found to be dangerous in formation flying.

Shortly after the return from the Middle East, the 363rd and its squadrons dropped the 'Tactical' from the designation. As one of 9th AF's premier units the Shaw wing has returned often to man the USAF detachment in Saudi Arabia, 9th AF being the designated air component of US Central Command. During one of the detachments, a 33rd FS F-16D flying from Dhahran scored the first kill by a USAF F-16, and the first by AIM-120 AMRAAM, when Lt Col Gary North shot down an Iraqi MiG-25 on 27 December 1992 over southern Iraq. The 363rd had only just converted to the F-16C/D Block 42. These aircraft were not in service for long, as the wing then converted to the latest 'Viper' variant. Beginning with the 78th FS, the Shaw wing began equipping with the Block 50D version, compatible with the ASQ-213 HARM Targetting System pod to give defence suppression capability. By 1995 all three squadrons had HTS-compatible aircraft, although the anti-radar role is a secondary mission.

In January 1994, as part of a major reshuffle, the 363rd Fighter Wing and its constituent squadrons were renumbered, the wing inheriting the 20th Fighter Wing numberplate from the deactivated F-111E wing in England. The three squadrons also renumbered, becoming the 77th FS 'Gamblers', 78th FS 'Bushmasters' and 79th FS 'Tigers'. The 77th and 79th had previously been F-111E squadrons at RAF Upper Heyford, while the 78th number came from the 81st TFW with A-10s at RAF Woodbridge. The third Upper Heyford squadron number, the 55th FS, was assigned to Shaw's single A-10 squadron. The three F-16 squadrons changed their markings accordingly, the 77th applying its nickname on a red stripe with two ace-of-spades, the 78th featuring a red snake on a yellow fin band while the 79th features a tiger-striped fin-band. The wing commander's aircraft features a multi-coloured fin-stripe, and the 20th FW also provides an aircraft marked for the commander of the 9th Air Force.

Above: The 79th FS operates Block 50 aircraft with HTS equipment.

Right: In addition to the three F-16 squadrons, the 20th FW operates a single squadron (55th FS) of A-10s.

Below: A 'Bushmasters' HTS aircraft leads a 'Gamblers' machine loaded out with Mk 84s. All three squadrons have HARM-capable aircraft.

23rd Wing
Pope AFB, North Carolina 'FT'

74th Fighter Squadron (blue/white)

After a period of inactivity, the 23rd was reactivated in 1964 to operate the F-105D. A-7s were received in 1972, and A-10s in 1980. From England AFB, Louisiana, the 23rd dispatched two squadrons to Desert Storm.

On 1 June 1992 the 'old' 23rd Fighter Wing (at England with 'EL' tailcode) deactivated and the 'new' 23rd Wing stood up at Pope AFB, North Carolina. This was the first of the composite wings dedicated to the support of a major army unit, in this case the 82nd Airborne Division at nearby Fort Bragg. One of the 23rd FW's original A-10 units (75th FS) was reactivated on 1 April (prior to the wing change) as the first unit, and two C-130E squadrons were added to the new wing (2nd ALS and 41st ALS). Completing the wing line-up was the addition of an Fighting Falcon squadron in the form of the 74th Fighter Squadron 'Flying Tigers'. The 23rd Wing adopted the traditions of the Flying Tigers, and all aircraft wear the 'FT' tailcode and sharksmouth. The F-16s wear a blue fin-band with white stars and the legend '74th' superimposed. One aircraft is marked for the wing commander with a multi-coloured fin-band.

The colourful F-16s of the 74th FS are employed primarily on CAS duties in support of the 82nd Airborne. The 'boss bird' (right) is seen carrying a LANTIRN pod.

They are Block 40 Night Falcon aircraft, tailored for the night CAS role with LANTIRN pods.

In 1994 the 23rd Wing was the lead unit in the planned invasion of Haiti, and also took part in the major reinforcement of Kuwait in response to Iraqi manoeuvring near the Kuwaiti border.

31st Fighter Wing
(deactivated)

Homestead AFB, Florida 'ZF'/'HS'

306th Fighter Squadron
(yellow - not current)
307th Fighter Squadron
(red - not current)
308th Fighter Squadron
(green - not current)
309th Fighter Squadron
(blue - not current)

Returning from Vietnam in October 1970, the 31st TFW operated F-4s in the fighter-bomber role until 1 October 1982, when it received its first F-16A/B Block 15. At the time the tailcode was 'ZF', but this changed to 'HS' in 1985 to reflect the base name. The 306th TFS deactivated leaving the wing with three F-16 squadrons flying missions in the strategically important Florida region close to Cuba. The three squadrons carried coloured fin-bands with the squadron name superimposed. The 307th was the 'Stingers', the 308th the 'Knights' and the 309th the 'Wild Ducks'. During 1990 the wing converted to the F-16C/D Block 40, but was deactivated in 1993, its squadron numbers resurfacing at Luke as F-16 training units, and the wing number transferred to USAFE's new wing in Italy.

In the early 1990s the 31st Fighter Wing was flying the Block 40 with three squadrons. After deactivation two of the squadrons were transferred to Moody, while the wing number was resurrected at Aviano in Italy.

56th Fighter Wing
(deactivated)

MacDill AFB, Florida 'MC'

61st Fighter Squadron
(yellow - not current)
62nd Fighter Squadron
(blue - not current)
63rd Fighter Squadron
(red - not current)
72nd Fighter Squadron
(black - not current)

Inheriting the numberplate of a special operations wing, the 56th TFW was established at MacDill in 1975 to operate F-4Es. On 22 October 1979 the unit became the second USAF unit to acquire Fighting Falcons in the form of the F-16A/B Block 10. Aircraft wore an 'MC' tailcode, and carried a coloured fin-band according to squadron. The 56th was established as an F-16 training unit, and in July 1982 was redesignated as a Tactical Training Wing. A variety of both Pratt- and GE-powered aircraft were on charge until the wing standardised on the F-16C/D Block 30 in

1990. In 1991 the wing and squadrons were redesignated as FW and FS respectively but, in 1994, the training task was consolidated with the 58th FW at Luke. The squadron numbers and wing numberplate were then assigned to the Luke unit.

Initial equipment for the 56th TTW was the F-16A/B Block 10 (below left). Later the F-16C/D Block 30 (below) was introduced. Both are from the 61st TFTS 'Top Dawg'.

347th Wing

Moody AFB, Georgia 'MY'

68th Fighter Squadron (red)
69th Fighter Squadron (silver)
70th Fighter Squadron
(blue/white – not current)
307th Fighter Squadron
(red – not current)
308th Fighter Squadron
(green – not current)

Following a spell operating F-111As in Southeast Asia, the 347th TFW reactivated at its present base at Moody with F-4Es. These were flown until 1988 when the 68th, 69th and 70th TFS converted to the F-16A/B Block 15, all adorned with the 'MY'

tailcode and fin-bands in the squadron colours. The wing soon updated to F-16C/D Block 40 aircraft with LANTIRN pods, and dispatched the 69th TFS for Desert Storm operations. Flying from Alminhad, United Arab Emirates, the 69th operated as part of the 388th TFW(P). In 1993 the wing absorbed the 307th and 308th FS from the deactivating 31st FW in preparation for a major change to become the second army support composite wing. Designated simply 347th Wing on 1 July 1994, the Moody unit lost the 69th, 70th and 308th FS F-16s, while the 307th renumbered as the 'new' 69th FS. The 70th FS number was transferred to a new A-10 squadron, while the 52nd AS was added with C-130Es. In the new look wing, the 347th now has the 68th and 69th FS flying the Block 40.

Now a composite air-land wing with A-10s and C-130s in addition to F-16s, the 347th briefly had the 308th FS (above) under its control following deactivation of the 31st FW. The two current 'Viper' units are the 68th FS (below left) and 69th FS (below), both of which operate the F-16C/D Block 40 with LANTIRN on close air support/air interdiction duties.

12th Air Force

366th Wing

Mountain Home AFB, Idaho 'MO'

389th Fighter Squadron (red)

Operating F-111s since 1976, the 366th 'Gunfighters' and its Mountain Home base were chosen to be the unit and location of the USAF's power projection 'super wing'. The unit was redesignated the 366th Wing on 1 January 1992 and began shipping out its 'Aardvarks' that spring. F-16C/D Block 25s arrived in the summer to adopt the 389th FS 'Thunderbolts' number from one of the F-111 squadrons. They were subsequently joined by the 390th FS (F-15C), 391st FS (F-15E), 22nd ARS (KC-135R) and 34th BS (B-52G, now B-1B) to complete the 'mini air force'.

After some initial operational problems, the 366th undertook several joint training exercises, including a 1993 Green Flag, trips to Alaska and Egypt, and a 1995 Red Flag. At the end of 1994 the 389th upgraded from the Block 25 to the Block 52D variant of the F-16, Pratt-engined for commonality with the wing's F-15s and compatible with the ASQ-213 pod and AGM-88 HARM missile for providing a defence suppression capability. The F-16s wear the 'MO'

tailcode, with a red fin-band containing a yellow thunderbolt.

The single F-16 squadron within the 366th is tasked with general attack (this pair carrying Mavericks) and with defence suppression duties.

388th Fighter Wing

Hill AFB, Utah 'HL'

4th Fighter Squadron (yellow)
16th Tactical Fighter Training
Squadron (blue/white – not current)
34th Fighter Squadron (red)
421st Fighter Squadron (black)

Based at Hill, the 388th was not only the first operational recipient of the F-16, but has also operated the most variants. Having been one of the main combat formations in Southeast Asia, the wing set up shop at Hill in December 1975, operating F-4Ds. On 6 January 1979 it made history by receiving

Hill was the first base to receive the F-16, the initial 388th TFW examples being Block 1 machines. These were easily identifiable thanks to the black radome.

Between 1983 and 1989 the 388th operated the Block 15 F-16A as its principal variant.

the first operational F-16, delivered to the 16th TFTS. Initial deliveries consisted of aircraft from Blocks 1, 5 and 10, and were

Block 40 aircraft arrived from 1989, and these were dispatched to the Gulf. This D model is from the 4th Fighter Squadron.

used to rapidly train pilots, including instructors from the initial four European purchasers. The wing delcared IOC in October 1980, and made the first overseas USAF deployment in March 1981 to Norway. In 1983 the wing scored the F-16's first Gunsmoke victory, and during the year it upgraded to the Block 15 aircraft.

In 1987 the 16th TFTS was deactivated, leaving the 388th TFW as a three-squadron unit. On 17 May 1989 the wing received the first of its F-16C/D Block 40 LANTIRN-

capable aircraft. Two squadrons of these, the 4th and 421st FS, were sent to Alminhad AB in the United Arab Emirates to form the core of the 388th Tactical Fighter Wing (Provisional), joined by another LANTIRN squadron from Moody during Desert Storm operations. During 1993 the 4th FS began to receive Block 50 aircraft. That year the wing won Gunsmoke again. All 388th aircraft have worn the 'HL' tailcode and squadron colours on the fin-bands (including a multi-coloured aircraft for

the wing boss). The 4th FS adds a small lightning bolt while the 34th FS 'Rams' and 421st FS 'Black Widows' also wear the squadron nickname. The 34th and 421st did not upgrade to the Block 50 early 1995 the 4th reverted to Block 40 aircraft. That year Hill was also named as a possible candidate for base closure.

The 4th FS was the first recipient of the Block 50, although it subsequently returned to Block 40s.

474th Tactical Fighter Wing (deactivated)

Nellis hosted an operational wing of Block 10 F-16s during the 1980s, but this was deactivated as the first of many victims of budgetary cuts.

Nellis AFB, Nevada 'NA'

428th Tactical Fighter Squadron
(blue – not current)
429th Tactical Fighter Squadron
(yellow – not current)
430th Tactical Fighter Squadron
(red – not current)

Previously operating the F-111 and F-4 (the latter in the training role), the 474th TFW at Nellis was chosen as the third wing to receive F-16s, which began arriving on 14 November 1980. Initially these were Block 1 and 5 aircraft, but subsequently Block 10s were used. Aircraft all carried the 'NA' tailcode, and fin-bands relating to the squadron colour. The 428th 'Buccaneers' also carried a skull-and-cross-bones, the 429th 'Black Falcons' a small bird of prey, while the 430th 'Tigers' featured a red/white chevron fin-stripe. A restructuring of tactical assets in September 1989 resulted in the 474th TFW being inactivated and its aircraft redistributed. The three squadrons resurfaced as F-111 operators under the 27th FW.

USAF Weapons and Tactics Center

57th Wing

Nellis AFB, Nevada 'WA'
(yellow/black)

64th Agressor Squadron
  (not current)
414th Training Squadron
  (Adversary Tactics Division)
422nd Test and Evaluation Squadron
F-16 Fighter Weapons School (WSF)
Aerial Demonstration Squadron
  ('Thunderbirds')

What is now the USAF WTC was
established at Nellis in 1966 as the Tactical
Fighter Weapons Center, with the 4525th
Fighter Weapons Wing as its flying unit. The
role of the unit was to test and evaluate
new weapons and tactics for the tactical
fighter force, and to provide advanced
tactical instruction to experienced squadron
aircrew. In October 1969 the unit
renumbered as the 57th FWW. The
designation has changed several times in
the ensuing years, but since 1994 has been
the 57th Wing to reflect the wide nature of
the unit's tasks. Aircraft of the WSF and
422nd TES wear a 'WA' tailcode and a
yellow/black fin-stripe.

**Above: The 57th operates
grey F-16s in both
advanced tactics training
and test roles. This is a
Block 42 machine with
LANTIRN pods, operated
by the 422nd TES.**

**Below: In common with
the other Nellis-based
F-16s, the 'Thunderbirds'
operate P&W-powered
aircraft. They are now
equipped with F-16C
Block 32s.**

The wing sent a detachment to Hill AFB
in 1980 to take part in the MOTE
programme which provided a cadre of
instructors. The first F-16A Block 1/5/10s
arrived at Nellis in September for the 422nd
Test and Evaluation Squadron, which is still
the main trials squadron within the wing.
The 422nd subsequently established two
detachments, Det 2 at Luke AFB and Det
16 at Hill AFB. Small numbers of F-16A/B
Block 15s and F-16C/D Block 25s were
used, and on 20 June 1987 the wing
received its first F-16C/D Block 32. This
variant is still the predominant variant at
Nellis, although the wing now has Block 42
and 52 aircraft.

While the 422nd TES handles the
evaluation duties, the 57th has the F-16
Fighter Weapons School (WSF – Weapons
School, Falcon) for instruction purposes.
Pilots attend the WSF to undertake a highly
advanced course in tactics, following which
they return to their units to impart their
new-found knowledge. The 57th FWW
previously operated two squadrons of
aggressor aircraft equipped with the F-5E.
One of these, the 64th AS, subsequently
converted to the Block 32 F-16C/D, but with
the cancellation of the USAF's adversary
programme, the 64th was deactivated on 5
October 1990. An aggressor tradition is
maintained by a small cadre of pilots with
nine F-16C/D Block 32s, suitably painted in
Russian-style camouflage, which serve with
the Adversary Tactics Division of the 414th
Training Squadron (formerly 4440th TFTG),
the unit responsible for organising the Red
Flag exercises.

The final F-16 operating unit at Nellis is
by far the most famous: the Aerial

**Above: The 'Thunderbirds' have
used the F-16 for their immaculate
formation displays since 1983. F-16A
Block 15s were used initially.**

Demonstration Squadron. The
'Thunderbirds' transitioned to the F-16A/B
Block 15 in 1985, upgrading to the F-16C/D
Block 32 in 1993. The unit is away for much
of the year, but spends time in the winter at
Nellis with display work-up undertaken
using the runway at Indian Springs AFAF.

*Although the USAF's aggressor
programme is officially dead, the
414th TS/AT retains a cadre of
experienced pilots and a handful of
suitably decorated aircraft for Red
Flag adversary work.*

USAF Air Warfare Center

79th Test and Evaluation Group

*The 'OT' tailcode of the 79th TES stands for 'Operational Test', and from its
Eglin base the unit conducts a wide range of trials missions of different
equipment. This Block 42 aircraft carries a non-standard EW fairing on the
intake side similar to that carried by Belgian and Israeli aircraft.*

Eglin AFB, Florida 'OT'

85th Test and Evaluation Squadron
(black/white)

In business since 1963 (its designation
prefixed by 'Tactical' until 1 June 1992), the
Air Warfare Center at Eglin AFB has
operated a single flying squadron on trials
and evaluation duties, chiefly concerned
with weapon systems. On 2 October 1992
this unit, the 4485th Test Squadron, was
redesignated as the 85th TS, and is now
referred to as the 85th TES. The unit has
operated many F-16s on test duties since
the early days of the programme, including
Block 10, 15, 25, 32, 42 and 50 aircraft. A
recent major programme has been the rapid
evaluation for service use of the ASQ-213
HARM Targetting System using Block 50D
aircraft.

Air Education and Training Command

2nd Air Force

82nd Training Wing   Sheppard AFB, Texas

Formerly the 3700th Technical Training Wing, the 82nd TW provides technical instruction for ground crew. A large number of ground instructional airframes are maintained at the base for this purpose, all assigned the 'G' prefix to denote their non-flying status. Examples of the GF-16A, GF-16B and GF-16C are on strength.

19th Air Force

56th Fighter Wing
(formerly 58th Fighter Wing)

Luke AFB, Arizona 'LF'

61st Fighter Squadron
   (yellow/black)
62nd Fighter Squadron (white/blue)
63rd Fighter Squadron (red/black)
308th Fighter Squadron (green/white)
309th Fighter Squadron (blue/white)
310th Fighter Squadron
   (green/yellow)
311th Fighter Squadron (blue/white)
312th Tactical Fighter Training
   Squadron (green/orange – not
   current)
314th Fighter Squadron
   (yellow/black – not current)
425th Fighter Squadron (black/red)

Situated to the west of Phoenix, Luke AFB offers unrivalled flying weather and nearby weapon ranges – perfect for fighter training. The 58th Tactical Fighter Training Wing was established in October 1969 by the renumbering of the 4510th CCTW, and subsequently trained pilots for the A-7, F-100, F-4, F-5 and F-15. In 1977 it was redesignated as a Tactical Training Wing.

In February 1983 it became the second major F-16 training unit with the delivery of its first F-16A/B Block 10 aircraft. At that time the tailcode changed from 'LA' to 'LF' (for 'Luke Falcon'), and later the Block 15 was added. Initially there were three

All active-duty F-16 training is concentrated at Luke, with eight squadrons operational. The 310th has survived since the early days.

squadrons: the 310th TFTS 'Top Hats', 311th TFTS 'Sidewinders' and 312th TFTS 'Scorpions', but the 314th TFTS 'Warhawks' was added on 3 October 1986. By that time the wing was in the process of converting to the F-16C/D Block 25, with the exception of the 311th TFTS which remained on F-16As to train foreign pilots. For a brief period the 58th TTW simultaneously operated the first production examples of the F-16A, B, C and D.

On 1 October 1991 the wing became the 58th Fighter Wing and, having lost the 312th TFTS previously, added two squadrons of F-15Es to consolidate all Luke-based training under one wing. By March 1995 all F-15Es had left, the training commitment having been passed to Seymour Johnson AFB.

In the interim, the F-16 squadrons had been undergoing a major reshuffle in an attempt to keep alive famous units from disbanding wings elsewhere. The 58th also changed from Air Combat Command to AETC, which formed on 1 July 1993. At the heart of the squadron reorganisation was the closure of MacDill and the transfer of its F-16 training commitment to Luke. On 1 April 1994 the 58th Fighter Wing assumed the MacDill wing numberplate, becoming the 56th Fighter Wing. MacDill's 61st, 62nd and 63rd squadron numbers were also assigned to Luke. The 308th and 309th numbers came from the ex-Homestead 31st FW. The 425th FS is expected to adopt the 307th FS number in the future. Only the 310th and 311th FS from the original 58th TTW lineup have survived.

Thus the 56th FW now boasts eight F-16 squadrons, standardised on the Block 42 with the exception of the 61st and 62nd (retaining Block 25s) and the 425th, which retains Block 15 A/Bs for the training of FMS (foreign military sales) pilots. This mission is to be handed over to the Arizona ANG in late 1995, with the 425th receiving F-16C/Ds. A new squadron is expected to train Taiwanese F-16 pilots.

Above: the F-16C/D Block 42 is the standard Luke variant, this example being from the 63rd FS.

Below: At the outset the 58th TTW was equipped with F-16A/B, initially Block 10s (illustrated).

Air Force Materiel Command

Air Force Development Test Center

46th Test Wing
(formerly 3246th Test Wing)

Eglin AFB, Florida 'ET' (red/white)

39th Flight Test Squadron
40th Flight Test Squadron
3247th Test Squadron (not current)

Responsible for the development and testing of weapons, the 3246th Test Wing was established in July 1970 as the flying component of the Air Proving Ground Center (now known as the Air Force Development Test Center, Munitions Systems Division). Its flying squadron was the 3247th Test Squadron. The first F-16A/B Block 5s were assigned to the 3247th in 1980 just as the type was entering service.

The work of the squadron was to evaluate the launch envelope for each weapon and clear them for use.

Aircraft were initially marked with an 'AD' tailcode (for 'Armament Division'), but on 30 September 1989 this changed to 'ET' for 'Eglin Test'. The fin-stripe is white with red diamonds. On 2 October 1992 the wing renumbered as the 46th Test Wing, the 3247th TS becoming the 40th FTS.

Subsequently the 39th FTS was added, this unit conducting missile tests. F-16s have been assigned in small numbers, covering the major variants, including the Block 15 ADF. In 1995 the 39th FTS conducted captive-carry tests of the BAe ASRAAM missile using an F-16C Block 25.

This 39th FTS F-16 was used by the Eglin armament test unit for captive-carry tests of the BAe Dynamics ASRAAM short-range air-to-air missile.

Previously the 3247th TS, the 40th FTS wears red diamonds on the fin, as opposed to blue for the 39th. This Block 40 aircraft carries a pair of GBU-12s.

Air Force Flight Test Center

412th Test Wing
(formerly 6510th Test Wing)

Edwards AFB, California 'ED'

416th Flight Test Squadron
  (blue/white)
6512th Test Squadron
  (blue/white – not current)

In March 1978 the 6510th TW was formalised as the flying component of the Air Force Flight Test Center, headquartered at Edwards AFB. As such, it and its flying squadron, the 6512th TS, conducted a wide range of test and evaluation missions, operating at one time or another virtually every type in the USAF inventory. Edwards was the site for the YF-16/YF-17 fly-off, and was also the main test location of the FSD work with the pre-production F-16s. Since that time many F-16s have been assigned to Edwards, covering all major variants of the type.

On 1 October 1992 the 6510th TW was renumbered as the 412th Test Wing, and its disparate fleet was organised into squadrons by type. All F-16 flying is now conducted by the 416th FTS. In addition to the ongoing trials work of new variants and new equipment, the 416th also provides aircraft for chase work and for the Test Pilot School. Several elderly Block 10/15 aircraft are retained for these general-purpose duties, mostly painted in a high-conspicuity red/white scheme. Many aircraft of the Edwards trials fleet have been upgraded with equipment associated with later

Aircraft assigned for chase or test pilot duties (mostly Block 10 F-16A/Bs) wear a high conspicuity scheme. This example is used for B-2 chase duties.

variants. Recent major trials programmes have included the evaluation of the F-16ES with overwing conformal tanks, automatic target hand-off systems and targeting by satellites. The 416th also jointly operates the NF-16A AFTI aircraft. All aircraft carry the 'ED' tailcode and blue fin-stripe with small white crosses.

The 412th conducts most of the F-16's aerodynamic trials. This aircraft was used for the Lockheed-funded ES trials in 1994.

Ogden Air Logistics Center

75th Air Base Wing

Hill AFB, Utah 'HAFB'

Confirming Hill AFB's claim as the capital of the F-16 community is the co-located Ogden Air Logistics Center, which undertakes all F-16 depot-level maintenance. Although not officially an aircraft-owning agency, the OALC nevertheless has generated enough of its own test work to justify the use of a single F-16A Block 10 aircraft (79-0402) permanently assigned to the depot. Later an F-16B Block 15 (81-0817) was added. These are used to evaluate new

maintenance procedures and modifications devised by the depot. It wears the same high-conspicuity scheme as the Edwards aircraft, but carries the letters 'HAFB' on the fin (for Hill AFB). After many years without such organisation, AFMC established during 1994 numbered Air Base Wings to control flying activities at its bases, Ogden becoming the 75th ABW. In 1995 Hill and the associated ALC were high on the list for potential closure.

The Ogden Air Logistics Center operates a pair of F-16s (one A and one B) for test duties. These evaluate new procedures and modifications produced by the ALC itself.

Pacific Air Forces

5th Air Force

18th Tactical Fighter Wing

(deactivated)

Kadena AB, Japan 'ZZ'

26th Aggressor Squadron (not current)

During the 1980s the 26th AS was PACAF's resident Aggressor unit, flying F-5Es from Clark AB, Philippines. Five F-16C Block 30s

were dispatched to the Far East on 29 August 1989 to form the new equipment of the squadron, which was in the process of moving to Kadena AB on Okinawa. The F-16s were temporarily assigned to Osan in the standard fighter role while the 26th AS completed its move. However, the USAF decision to dismantle its Aggressor

programme before the 26th AS was ready to begin operations resulted in the F-16s being absorbed into the Osan unit. If they had have reached Kadena, they would have been the only USAF Aggressors to be powered by the General Electric F110 engine.

35th Fighter Wing

(formerly 432nd Fighter Wing)

Misawa AB, Japan 'MJ'

13th Fighter Squadron (red)
14th Fighter Squadron (yellow)

The 432nd Tactical Reconnaissance Wing was one of the major units during the Southeast Asia war, and after a spell operating drones in the US was reactivated in July 1984 at Misawa AB in northern Honshu, Japan, to become the second PACAF F-16 wing. The initial equipment for the two squadrons was the F-16A/B Block 15. In terms of markings the aircraft wore an 'MJ' tailcode (for 'Misawa, Japan'), the 13th TFS featuring a black/white checkerboard and the 14th TFS a yellow/black checkerboard. These markings were transferred to the F-16C/D Block 30s which arrived in 1990, but were soon changed to a simple red band for the 13th and yellow for the 14th. During 1992 the wing renumbered as the 35th Fighter Wing, adopting the numberplate from the disbanding F-4 wing at George AFB. In earsly 1995 the 14th FS received HTS-equipped Block 50 aircraft.

Below: Prior to the current markings the Misawa aircraft wore a checkerboard fin-stripe.

Above: A 14th FS Block 30 aircraft displays the wingtip AIM-120 AMRAAM mounting.

Below: When the 432nd FW renumbered as the 35th the squadrons remained unchanged.

7th Air Force

8th Fighter Wing

Kunsan AB, Republic of Korea 'WP'

35th Fighter Squadron (blue)
80th Fighter Squadron (yellow)

Partly as compensation for a considerable drop in US troop numbers in Korea, the 8th Tactical Fighter Wing at Kunsan AB was chosen to be the first overseas unit (fourth overall) to receive the Fighting Falcon. The 'Wolf Pack', as the wing is known, had spent most of its USAF career stationed in the Pacific theatre, including combat in Korea and Vietnam. After the end of the war

When it first received F-16s, the 8th TFW operated the Block 10. These were supplanted by Block 15 aircraft (illustrated). The wolf's head motif is carried behind the cockpit.

in Southeast Asia, in which it was the highest-scoring air-to-air unit, the wing relocated to Kunsan with the F-4D.

On 15 September 1981 the first F-16A/B Block 10s arrived, followed later by Block 15 aircraft. Aircraft were marked with the coloured fin-bands of the two squadrons (35th TFS 'Phantoms' and 80th TFS 'Juvats'), and carried a 'WP' tailcode and a wolf's head. In 1987 the F-16C/D Block 30 arrived to improve the wing's capability, later bolstered by the transfer of aircraft from the 51st TFW. These aircraft are still in service in a dual-role mission, although the wing is slanted to an air defence role. In late 1991 the wing was redesignated as a Fighter Wing.

Current 8th FW C models wear the wolf's head badge on the fin. This 35th FS aircraft is from the very last batch of Block 30 aircraft produced for the USAF. It is armed with Mk 82 LDGP bombs in a 'slant-two' arrangement, and carries an ALQ-184 (long) ECM pod on the centreline. Owing to continuing friction in the Korean peninsula, the 8th and 51st FWs maintain a high state of readiness.

51st Fighter Wing

Osan AB, Republic of Korea 'OS'

36th Fighter Squadron (red/black)

One of two main fighter wings during the Korean war, the 51st remained in the Pacific theatre at various bases in Japan and Korea. After moving to Osan in the early 1970s it started operating the F-4E, A-10 and OV-10, and, in addition to the headquarters base, it maintained detachments at Suwon and Taegu. In 1989 a single squadron converted to the F-16C Block 30 at Osan, while a year

later an OA-10 unit (19th TASS) joined the wing, moving into Osan from Suwon. Aircraft wear an 'OS' tailcode and a horse insignia, while the 36th TFS aircraft adopted a red/black checkerboard fin-stripe. In June 1990 the first F-16C/D Block 42 aircraft arrived – the only Pratt-engined C models sent overseas – and the Block 30s were passed on to the 8th TFW at Kunsan. On 1 October 1990 the 51st TFW became the 51st FW.

The 36th FS is principally a CAS unit, and operates LANTIRN-equipped Block 42 machines.

11th Air Force
354th Fighter Wing

(formerly 343rd Fighter Wing)

Eielson AFB, Alaska 'AK'

18th Fighter Squadron (blue)

The 343rd was established as a Composite Wing in 1981 with the A-10A and O-2A. OV-10s replaced the Cessnas in 1986, by which time the wing had redesignated as a TFW.

Like the 51st FW, the 354th operates a squadron of F-16C/D Night Falcons alongside a squadron of OA-10s which act in the FAC role.

In 1990 the Bronco was retired, leaving the 18th TFS 'Blue Foxes' as the only flying unit. This gained a few OA-10s before switching to the F-16C/D Block 40 in 1991. In March that year the unit became simply a Wing. Some of the 'Warthogs' displaced by the arrival of the F-16s were used to form the 11th TASS. The F-16s are used primarily

in the close air support role, supported by the OA-10s acting as FAC aircraft. The LANTIRN pods give the F-16s excellent night capability, and the 18th FS was instrumental in developing night CAS tactics. On 1 August 1993 the unit inherited the numberplate of the 354th Fighter Wing, which disbanded as an A-10 unit at Myrtle Beach AFB. The F-16 squadron remained unaffected, but the 11th TASS was renumbered as the 355th FS. Both aircraft types wear an 'AK' tailcode, while the F-16s have a blue fin-stripe and a fox motif behind the cockpit.

18th FS F-16s are seen during a day bombing training mission. Although the 18th has a night CAS speciality, it still retains the day CAS, AI and battlefield air superiority missions.

United States Air Forces in Europe

3rd Air Force

81st Tactical Fighter Wing (deactivated)

RAF Bentwaters, England 'WR'

527th Aggressor Squadron (not current)

Having been resident at RAF Alconbury for several years, the 527th AS was USAFE's Aggressor squadron, using the Northrop F-5E to give dissimilar air combat practice to both USAFE aircraft and NATO allies. During its tenure at Alconbury it was assigned to the 10th TRW/TFW. In January 1989 the F-16C Block 32 was adopted for the Aggressor task, and the squadron moved to RAF Bentwaters to begin operations on the type. The squadron was assigned to the resident 81st TFW, which controlled four squadrons of A-10s at the twin bases of Bentwaters and Woodbridge. Unlike other dedicated Aggressor aircraft, the F-16s of the 527th AS retained standard two-tone grey camouflage, although Soviet-style two-digit codes were applied in red on the fuselage aft of the cockpit. After a short career with F-16s, the 527th AS was deactivated in 1990 when the USAF disbanded its entire Aggressor programme as a cost-saving measure. The 527th aircraft were dispersed to other units.

The 527th had only a short career in F-16s prior to disbandment of the unit and cancellation of the Aggressor programme.

16th Air Force
31st Fighter Wing

Aviano AB, Italy 'AV'

510th Fighter Squadron (purple)
555th Fighter Squadron (blue)

When the 401st TFW was deactivated in 1991, it was planned that the unit would resurface in Italy at Crotone AB. However, the new Italian wing was actually established at Aviano AB, and assumed the 31st numberplate from the disbanded Homestead unit. Aviano had been the major US base in Italy, and since 1992 had supported sizeable detachments for the Deny Flight operation over former Yugoslavia. The new wing activated on 11 April 1994, with 'AV' tailcode, using LANTIRN-compatible F-16C/D Block 40 aircraft relocated from Ramstein AB. The first squadron was created by renumbering the 526th FS with the famous 555th 'Triple Nickel' numberplate, previously an F-15E training squadron at Luke. The second squadron was the 512th FS transferred directly from Ramstein. This squadron then renumbered as the 510th FS 'Buzzards' at the end of February 1994, assuming the identity of a USAFE A-10 unit. Both squadrons wear an 'AV' tailcode and carry the squadron colours as fin-bands. The 555th carries the 'Triple Nickel' legend on the fin-stripe, and the squadron commander's aircraft has the head of a bald eagle. The wing commander's aircraft has blue tail markings and the wing's wyvern motif. Aircraft from the 31st FW were involved in the 21 November 1994 attack on the Krajina Serb airfield at Udbina, flying top cover for the raid. In 1995 the wing began using its F-16s in an airborne forward air control role over Bosnia. In June 1995 a 555th FS aircraft piloted by Captain Scott O'Grady was downed by a Serb SAM, resulting in a successful and well-publicised rescue by Marine helicopters.

Right: The two squadrons of the 31st are heavily involved in Operation Deny Flight over Bosnia, and indeed lost an aircraft to an SA-6 in June 1995. This 555th FS aircraft is in a typical weapons configuration for this theatre, with AIM-120s on the wingtip, AIM-9s on the outboard pylon, Mk 82 bomb inboard and ALQ-131 ECM pod.

Below: The 31st FW 'boss-bird' wears colourful tail markings. By May 1995 the wing's 510th FS had performed 355 Deny Flight missions and the 555th FS 565 missions.

401st Tactical Fighter Wing (deactivated)

Torrejon AB, Spain 'TJ'

612th Tactical Fighter Squadron
(blue/white – not current)
613th Tactical Fighter Squadron
(yellow/blue – not current)
614th Tactical Fighter Squadron
(red/blue – not current)

During the 1980s the 401st TFW's F-16s were the backbone of USAF power in southern Europe. F-16A/B Block 15s (right) gave way to F-16C/D Block 30s (below), which were used in Desert Storm. The wing's multi-role tasking embraced tactical nuclear delivery in addition to the standard fighter-bomber duties.

To bolster NATO's southern flank, the 401st TFW was issued with Fighting Falcons in April 1983 to replace F-4D Phantoms. Bearing the 'TJ' tailcode, the wing had three F-16 squadrons: the 612th 'Fighting Eagles', 613th 'Squids' and 614th 'Lucky Devils'. Squadron colours were worn as checkerboard fin-stripes, while a falcon's head and lightning bolt appeared on the tails of most aircraft. One aircraft was marked for the commander of 16th Air Force. Block 15A/Bs were the initial equipment, although the wing upgraded to the F-16C/D Block 30 in 1989. The 401st TFW had a multi-role mission, including air defence and nuclear delivery. As the only tactical unit within 16th Air Force, it regularly practised from operating at FOLs throughout the 16th AF region, including Aviano in Italy and Incirlik in Turkey. In 1989 the Spanish government voted to evict US combat units from its territory, and the 401st prepared to resurrect at Crotone AB in Italy. The Iraqi invasion of Kuwait intervened, and the 401st TFW dispatched its 614th TFS to Qatar to provide air defence in the Gulf region. The wing deactivated in 1991 after Desert Storm, and the new wing in Italy was eventually formed in 1993 at Aviano as the 31st FW.

17th Air Force
50th Tactical Fighter Wing (deactivated)

Hahn AB, West Germany 'HR'

10th Tactical Fighter Squadron
(blue – not current)
313th Tactical Fighter Squadron
(orange – not current)
496th Tactical Fighter Squadron
(yellow – not current)

Hahn first played host to the 50th in 1953, and after a spell in France the wing settled there permanently as one of the stalwarts of USAFE, flying F-100s, F-102s, F-104s and F-4s. In July 1982 the wing became the first USAFE wing (and the second overseas unit)

The 50th TFW became the second overseas F-16 recipient when it acquired Block 15 machines.

to receive the F-16, taking delivery of Block 15 aircraft for its three squadrons. Markings consisted of an 'HR' tailcode, squadron colour fin-band and a striking hawk motif, the latter carried originally behind the cockpit and later on the fin. The unit was given a multi-role tasking within the 4 ATAF area, and conducted both air-to-air and air-to-ground training sorties. It was a regular particiopant in NATO exercises, and in 1985 provided the individual Gunsmoke winner. In 1986 the unit upgraded to the F-16C/D Block 25, and four years later the 10th TFS was deployed to Al Dafra, Sharjah, as part of the 363rd TFW(P) to take part in Desert Storm. Soon after the return of the 10th TFS, the 50th TFW was the first major casualty of the dramatic force decline in Europe, and was deactivated in 1991.

Above: The Hahn wing was among the first units to get F-16Cs, flying these until its deactivation. This aircraft flew with the 313th TFS.

52nd Fighter Wing

Spangdahlem AB, Germany 'SP'

22nd Fighter Squadron (red)
23rd Fighter Squadron (blue)
81st Fighter Squadron
  (yellow – not current)
480th Fighter Squadron
  (red – not current)

In December 1971 the 52nd TFW was reactivated at Spangdahlem AB to operate Phantoms in the 'Wild Weasel' defence suppression role. Initial equipment consisted of F-4Cs and F-4Ds, subsequently changing to F-4Es and F-4Gs. From April 1987 the 52nd TFW began replacing its F-4Es with F-16C/D Block 30s, each of its three squadrons operating 12 Fighting Falcons alongside 12 F-4Gs, operating in a hunter-killer team. The F-16s were uniquely wired for the carriage of the AGM-45 Shrike anti-radiation missile. The Spangdahlem squadrons at the time of the F-16's arrival were the 23rd TFS (blue fin-band), 81st (yellow) and 480th TFS (red). In addition to the fin-band the aircraft carried an 'SP' tailcode and a sharksmouth. In 1991 the 23rd TFS was deployed to Incirlik in Turkey to form part of the 7440th Composite Wing which was attacking northern Iraq during Desert Storm. The mixed F-16/F-4G force provided defence suppression in the region.

Following Desert Storm the wing was rationalised into two F-16C units (23rd and 480th FS) and one of F-4Gs (81st). On 16 January 1993 a 23rd FS aircraft scored a kill against an Iraqi MiG-29 over northern Iraq during policing operations. The F-4Gs finally left in late 1993, and the 81st squadron number was passed to a newly acquired A-10 unit. The disbandment of the 36th TFW at nearby Bitburg also saw the 480th

FS renumbering as the 22nd FS to keep alive the number of the former F-15 unit. Another Bitburg squadron, the 53rd FS, also moved to the 52nd FW, completing the current Spangdahlem line-up.

During 1993 the wing changed to the F-16C/D Block 50, these aircraft being compatible with the AGM-88 HARM, and subsequently acquired aircraft with the ASQ-213 HTS defence suppression pod. The 52nd has been heavily involved in US operations over former Yugoslavia, with several detachments of its aircraft to Aviano for Deny Flight operations.

Above: Today the 52nd is a composite wing with two F-16 squadrons, one of A-10s and one of F-15Cs. The wing has been heavily committed to supporting the UN effort in Bosnia, these 23rd FS aircraft being seen on a Deny Flight mission in 1993 when the squadron flew 404 combat missions.

Below: When the 52nd first took the F-16 on charge it was issued to the three constituent squadrons to fly alongside the F-4G in a hunter-killer defence suppression team. This pair is from the 480th FS.

The 81st FS was one of the three original 52nd TFW F-16 units before becoming the dedicated F-4G squadron. With the Phantoms departed, the 81st is now the number of the Spangdahlem A-10 squadron.

86th Fighter Wing
(not current)

Ramstein AB, Germany 'RS'

512th Fighter Squadron
  (green/black – not current)
526th Fighter Squadron
  (red/black – not current)

In Germany since its activation in 1948, the 86th flew F-84s, F-102s and F-4s. In December 1985 it received its first F-16s – Block 30 C/D aircraft – equipping the 512th TFS 'Dragons' and the 526th TFS 'Black Knights'. The initial tail markings consisted of the 'RS' tailcode and fin-stripe of alternating green/white bars for the 512th and red/black for the 526th. Dragon and knight badges were applied on the tails of the squadron commanders' aircraft. In late

1991 the wing became designated a Fighter Wing and in 1993/94 the two squadrons received Block 40 aircraft with LANTIRN capability. The tail markings changed subtly to a green or red fin-stripe outlined in black. These aircraft were operated only briefly before being assigned to the newly reactivated 31st FW at Aviano. The 86th FW had been heavily involved in manning the Deny Flight detachment at the Italian base. The 512th FS flew 783 missions between

14 September 1993 and 1 December, while the 526th flew 1,359 missions between 5 February and 28 June 1994. On 2 February 1994 two F-16s from the 526th FS shot down four Serbian aircraft, three with AIM-9s and one with AIM-120s. On 10 April F-16s bombed an artillery post near Gorazde. After losing its F-16 squadrons, the 86th is now an Airlift Wing, operating various transport types from Ramstein.

The 86th TFW was slated to receive three squadrons (plus the aggressors of the 527th AS), but in the event only two were assigned. The 'green-tails' were from the 512th TFS 'Dragons'.

The 526th squadron commander's aircraft wore this 'Black Knight' insignia on the fin. The unit was renumbered as the famous 'Triple Nickel', but not before a final blaze of glory when two F-16s shot down four Serbian aircraft.

Air Force Reserve

10th Air Force
301st Fighter Wing

Carswell Field, Texas 'TF'

457th Fighter Squadron

Reactivated in the Air Force Reserve at Carswell AFB in July 1972, the 301st TFW was initially equipped with F-105s, switching to F-4Ds in 1981, and the F-4E in 1987. These were flown briefly before the wing's 457th TFS adopted the Fighting Falcon in December 1990. The squadron, now designated 457th FS, flies the F-16C/D

Block 25 on fighter-bomber missions. The markings consist of the Texas flag at the top of the fin, the legend 'Fort Worth' and a longhorn steer's head at the base of the fin, and the tailcode 'TF' (for 'Texas Falcons'). Carswell closed as an active-duty AFB in 1993, but the 301st remained. In 1994 the unit celebrated its 50th anniversary, adding red and white rudder stripes to one aircraft. Between 1 December 1993 and 3 January 1994 the 301st FW manned a Deny Flight detachment at Aviano, flying 368 missions.

The 'Texas Falcons' of the 457th FS operate as a multi-role unit within the Reserve's 10th Air Force, which controls all AFRes Air Combat Command assets. The aircraft below wears a 50th anniversary scheme.

419th Fighter Wing

Hill AFB, UT 'HI'

466th Fighter Squadron

Apart from a brief existence as a Troop Carrier Wing in 1951, the 419th lay dormant until October 1982, when it was reactivated at Hill to operate F-105Ds in the attack role.

The career with the Thunderchief was short, for in January 1984 the wing's 466th TFS was chosen to be the first AFRes squadron to operate the Fighting Falcon. This was achieved by the simple expedient of transferring Block 1/5/10 aircraft from the co-located active-duty 388th TFW. The 466th's nickname is the 'Diamondbacks',

and in addition to the 'HI' tailcode the aircraft were given a diamond pattern fin-stripe in yellow and black, and a representation of a diamondback rattlesnake superimposed on the Utah outline. The 419th TFW excelled in its new machines, one of its pilots winning the 1987 Gunsmoke competition. In early 1994 the unit exchanged its elderly aircraft, some of the first F-16s built, for Block 30s. The markings remain similar, with the addition

of the nickname at the base of the fin. After a spell in Turkey on Provide Comfort missions, 419th FW aircraft were sent to Aviano for Deny Flight operations, crewed by personnel from the 704th and 706th FS.

The 466th TFS switched from the cumbersome F-105 to the nimble F-16A/B Block 10 in 1984. The fin-stripe was derived from the 'Diamondbacks' nickname.

A 419th FW F-16C lands at RAF Coltishall during a transit flight to Turkey for a Provide Comfort detachment. The aircraft carries an ALQ-131 ECM pod.

482nd Fighter Wing

Homestead ARB, Florida 'FM'

93rd Fighter Squadron

Brief periods of activation were followed in April 1981 by the activation of the 482nd TFW at Homestead AFB to operate the F-4C. The F-4D was operated from September 1983, and in July 1989 the wing's 93rd TFS adopted the F-16A Block 15 for dual-role operations. The aircraft were marked with a checkerboard fin-band,

The F-16s of the 'Florida Makos' are naturally marked with a shark.

'FM' tailcode (for Florida Makos) and the unit's nickname on the base of the fin. In addition a shark motif was applied to the fin.

In 1992 Homestead was hit by Hurricane Andrew, and the 93rd FS moved out to MacDill. It returned to Homestead in mid-1994 following renovation to the base, although in 1995 the field was under review for possible closure.

With their A/B Block 15 machines, the 93rd FS operates among the oldest F-16s in the USAF. The unit may be deactivated if Homestead is finally closed.

507th Fighter Group (deactivated)

Tinker AFB, Oklahoma 'SH'

465th Fighter Squadron (not current)

Based at Tinker AFB, the 507th TFG was administered by the 419th TFW at Hill, and was established in 1972 to operate F-105Ds with the 465th TFS, although these were retired the following year. In October 1980 the unit reactivated with the F-4D, and exactly eight years later began re-equipping with the F-16A/B Block 10 with a multi-role tasking. The aircraft carried a blue fin-band with the Oklahoma state seal, and were adorned with an 'SH' tailcode (for 'Shit Hot'). With the general downsizing of the force in the 1990s, the 507th FG was disbanded in 1994, the 465th becoming an Air Refueling Squadron with KC-135Rs.

10th Air Force had to lose two of its F-16 units during 1994, and the 465th FS was one of the unlucky pair. Operations at its Tinker base were becoming congested after the Navy's E-6 force took up residence.

906th Fighter Group (deactivated)

Wright-Patterson AFB, Ohio 'DO'

89th Fighter Squadron (not current)

After a career as a transport unit, the 906th TFG was reformed in July 1982 with the F-4D. In October 1989 the group's 89th TFS began the conversion to the F-16A/B Block 10, with which it undertook fighter-bomber duties from Wright-Patterson AFB. Aircraft carried a 'DO' tailcode (for Dayton, Ohio), and featured an orange/black or red/black checkerboard. A tiger's head was also carried on the fin, but this was changed to a snorting rhinoceros. As part of the cuts imposed on 10th Air Force, the unit was officially deactivated on 30 September 1994, although its last aircraft was flown out on 26 July. The 89th number was transferred to an Airlift Squadron flying C-141B StarLifters.

The 89th FS was one of the last USAF units with F-16A/B Block 10 aircraft, and was a natural choice for reroling as an airlift squadron (it now flies the StarLifter). Aircraft wore a rhino (below left) tail badge or a tiger (below).

924th Fighter Wing

Bergstrom ARS, Texas 'TX'

704th Fighter Squadron

Established as a transport unit in 1963, the 924th was inactive between 1976 and July 1981 when it was reformed as a fighter-bomber unit with F-4Ds, administered by the 482nd TFW at Homestead. After a brief period with F-4Es from 1989, the 924th TFG upgraded to the F-16A/B Block 15 in 1991. Wearing the Texas flag as a fin-stripe, the 'Outlaws' painted their nickname on the base of the fin, and applied the 'TX' tailcode. These markings were retained when the unit switched in 1994 to the Pratt-powered F-16C/D Block 32. In common with all AFRes units, the 924th FG was raised to wing status on 1 October 1994. From 14 February to 15 March 1995 the squadron manned a Deny Flight detachment at Aviano, using 419th FW aircraft. Combat sorties totalled 164 before the detachment was passed to the 926th FW.

The 301st has operated both the F-16A/B (below left) and F-16C/D (below). Both variants carried the 'Outlaws' nickname on the base of the fin.

926th Fighter Wing

NAS New Orleans, Louisiana 'NO'

706th Fighter Squadron

Activated in 1963 as the 926th TCG, the New Orleans-based Reserve unit was a transport unit until April 1978 when it converted to the Cessna A-37B. In January 1982 the 'Cajuns' of the 706th FS adopted the A-10A for the close support role, and the squadron took these to war in Desert Storm where they performed well (including a confirmed air-to-air kill). Soon after its return the 926th TFG dropped the 'tactical' from the designation in line with other Air Force Reserve units on 1 February 1992. By the end of the year they had re-equipped with the F-16C/D Block 30s to continue the CAS/BAI mission. Having previously been administered by the 917th TFW at Barksdale, the group was elevated to Fighter Wing status on 1 October 1994. Markings consist of an 'NO' tailcode (for New Orleans), the 'Cajuns' nickname on the base of the fin and yellow fleur-de-lys symbols on a red fin-band.

On 15 March 1995 the 706th FS took over the AFRES Deny Flight commitment, inheriting the 'HI'-coded F-16s previously flown by the 924th FW.

The 'Cajuns' of the 706th FS gained fame flying A-10s in Desert Storm, but in 1992 the squadron joined the growing ranks of F-16 users when it adopted Block 30 airframes. Like the other AFRes units it is assigned a battlefield role, and in March personnel were sent to Italy to take their turn on the Deny Flight effort.

944th Fighter Wing

Luke AFB, Arizona 'LR'

302nd Fighter Squadron

Having previously operated Sikorsky CH-3s, the 944th was reformed as a Tactical Fighter Group in June 1987 to operate the Fighting Falcon. It was the first unit in either the Reserve or ANG to be issued with factory-fresh F-16s. These were Pratt-powered Block 32s, and were used to win the 1989 Gunsmoke competition. They are still flown with a multi-role tasking. The group's flying squadron was the 302nd TFS 'Sun Devils', and their aircraft were accordingly marked with a small devil's head on an outline of Arizona, and a red/yellow fin-stripe depicting a trident. The 'LR' tailcode stands for 'Luke Reserve'. Two Reserve-wide edicts saw the designation changed to 302nd FS/944th FG on 1 February 1992, and 302nd FS/944th FW on 1 October 1994. From 3 January 1994 to 5 February the 302nd sent a detachment to Aviano for Deny Flight operations.

Although it was the first reserve unit to receive new F-16s rather than 'cascaded' secondhand machines, the 944th FG is still flying the same aircraft eight years later.

Air National Guard

107th Fighter Group, New York ANG (not current)

Niagara Falls IAP, New York

136th Fighter Squadron (not current)

The 136th FIS was committed to air defence of the United States between 1951 and 1960, and from 1970 when it transitioned to the F-101B Voodoo. During the 1980s it flew the F-4C and F-4D Phantom, undertaking its last F-4 operation on 31 August 1990. During that year the squadron began acquiring F-16A/B Block 15s to continue the air defence mission. These were subsequently all upgraded to ADF standard. Markings consisted of a representation of the Niagara Falls near the unit's base. On 15 March 1992 the 136th FIS was redesignated as a fighter squadron, and on 1 June 1992 the gaining command changed from TAC to ACC, although it remained essentially part of the 1st Air Force structure. In 1994, with no immediate threat being posed to the continental United States, the 136th FS gave up its F-16 air defence mission to become an air refuelling unit, equipped with KC-135Rs.

Assigned to the air defence mission, the 136th FS marked its aircraft with a tumbling falls motif. The squadron now applies a similar fin-stripe to its KC-135R tankers.

113th Fighter Wing, District of Columbia ANG

Andrews AFB, Maryland 'DC'

121st Fighter Squadron

Although part of the DC National Guard, the 121st FS flies from Andrews AFB in Maryland due to the lack of a suitable base within the District of Columbia. Since 1960 the 121st TFS has been a fighter-bomber unit, flying F-100s in the 1960s, F-105s in the 1970s and F-4Ds in the 1980s. In September 1989 the unit received its first F-16A Block 10 but it was not until the following fall that the conversion was completed. The squadron was redesignated 121st FS on 15 March 1992. In addition to the 'DC' tailcode, the aircraft wear a fin band with four white stars in a single-double-single arrangement to signify the unit's number. In mid-1994 the unit upgraded to the F-16C/D Block 30.

The DC ANG was one of the last USAF users of the Block 10 F-16A/B, distinguished by intake aerial.

114th Fighter Group, South Dakota ANG

Joe Foss Field, Sioux Falls, South Dakota

175th Fighter Squadron

Having spent the 1950s and 1960s as an interceptor unit, the 'Lobos' of the South Dakota ANG became a fighter-bomber unit when they adopted F-100Ds in 1970, subsequently operating the A-7D. With the Corsair they took part in Operation Just Cause, the December 1989 invasion of Panama. On 14 August 1991 the 175th received its first F-16C Block 30, but it was not until 30 January 1992 before the last Corsair was flown out to ther boneyard. With the F-16, the 175th (which became a Fighter Squadron on 15 March 1992) dispensed with the 'SD' tailcode worn by the A-7s and introduced a stylised wolf marking, with 'Lobos' on the fin-stripe and 'South Dakota' along the fun/fuselage fairing. In 1994 the 175th FS helped man an ANG deployment to Turkey for Operation Provide Comfort.

In addition to its individual nose art, this F-16C wears a Provide Comfort badge on the intake side to indicate a deployment to Incirlik in Turkey.

119th Fighter Group, North Dakota ANG

Hector Field, Fargo, North Dakota

178th Fighter Squadron

An interceptor unit guarding the northern United States since the early 1950s, the 178th FS continues to fly the mission today with F-16A/B Block 15 ADF aircraft. Conversion from the F-4D was accomplished during 1990, the last Phantom leaving Hector Field on 15 July. In addition to operations from home base, the 178th also mans air defence alerts at Klamath Falls, Oregon, and at March AFB, California.

State legislature allows the 178th FS to maintain its 'Happy Hooligans' nickname on the fins of its aircraft, the name having been earned during a particularly boisterous encampment. Also on the fin of the

commander's aircraft is a portrait of Teddy Roosevelt to commemorate the 1st Volunteer Cavalry Regiment, led by Roosevelt and composed largely of

Seen while on detachment to March AFB, the 178th 'boss-bird' carries the 'Rough Rider' portrait of Theodore Roosevelt not only on the aircraft's fuselage but also on the fin of the wingtip Sidewinder missile. The 'Happy Hooligans' are an integral part of the defence of the United States, and underlined their ability with a 1994 William Tell victory.

Dakotans, which fought bravely during the 1898 Spanish-American War. The Roosevelt portrait is carried behind the cockpit of regular squadron aircraft. In 1994 the self-appointed heirs to the 'Rough Rider' tradition scored a memorable victory at the William Tell air defence meet, notching up the first such victory for the F-16 in a competition normally dominated by F-15s.

F-16 Operators Part 1

120th Fighter Group, Montana ANG

Great Falls IAP, Montana

186th Fighter Squadron

The second of the air defence units to receive F-16s, the 186th FIS operated Convair F-106As until early 1987, when the first F-16A/B Block 15s were delivered. New fin markings were introduced consisting of the state name in a fin-stripe,

a steer's head superimposed on a stylised mountain scene, and the legend 'Big Sky Country' on the fin/fuselage fairing. The 186th FIS operated its F-16s in standard configuration until 1990/91, when they underwent conversion to ADF standard. The unit changed designation to 186th FS with all other Guard fighter squadrons on 15 March 1992.

Prior to ADF conversion, the 186th used regular F-16A/B Block 15s for air defence duties.

122nd Fighter Wing, Indiana ANG

Fort Wayne IAP, Indiana 'FW'

163rd Fighter Squadron

For years a fighter-bomber unit, the 163rd TFS was the ANG's last user of the F-4E Phantom, converting in the winter of 1991/92. The first of its F-16C/D Block 25s arrived in October 1991, while the last Phantom left on 21 January 1992. The 'FW' tailcode (for Fort Wayne) was carried across to the F-16s, as was the 'Indiana' fin-band. However, the F-16s also added the unit's nickname – 'Marksmen' – to the base of the fin, together with a rifle.

The 163rd FS was one of the last ANG units to convert to the F-16, acquiring Block 25s.

125th Fighter Group, Florida ANG

Jacksonville IAP, Florida 'FL'

159th Fighter Squadron

With an important task of defending the south-eastern corner of the United States, the 159th FIS has been an air defence unit since 1952. From 1974 to 1987 the squadron flew the F-106A Delta Dart, but in September 1986 it became the first of the ANG air defence units to adopt the F-16, although conversion to the F-16A/B Block

15 was not completed until 1 April 1988. In 1990 the aircraft were upgraded to ADF standard. The F-16s retained the FANG's blue fin-band with white lightning bolt, but later added an 'FL' tailcode. Based since 1968 at Jacksonville IAP, the 159th FS was expected to convert to the F-15A/B in 1995/96, still in the air defence role.

An F-16 of the 159th FS in full ADF standard lands during an exercise at Nellis, carrying an AIS pod on the wingtip. Note the repeat of the 'last-two' on the rear fuselage.

127th Fighter Wing, Michigan ANG

Selfridge ANGB, Michigan 'MI'

107th Fighter Squadron

During the 1960s the 107th was a tactical reconnaissance unit, but changed to a tactical fighter-attack squadron in 1972, flying F-100s. After a spell with A-7Ds the squadron converted to the F-16A/B Block 10, completing its re-equipment on 1 April 1990. Initially the aircraft carried a large falcon's head on the fin, with a checkerboard fin-stripe and the state name on the fin/fuselage fairing. Later the falcon's head was presented smaller and encircled. These markings were subsequently changed to a rainbow fin-stripe with the word 'Michigan' superimposed, and the bird replaced by the 'MI' tailcode. During 1994 the squadron upgraded to F-16C/D Block 30s.

Three generations of 107th F-16s: shown left is an F-16A wearing the initial markings, which were discarded in favour of a tailcode and fin-stripe (below). Today the markings have changed again, as has the equipment in the shape of the F-16C/D Block 30 (above).

128th Fighter Wing, Wisconsin ANG

Dane County RAP, Madison, Wisconsin 'WI'

176th Fighter Squadron

One of the last ANG units to convert to the F-16, the 176th FS had previously operated the A-10A in the close support role. The 'Warthogs' were shipped out between June and December 1992, while the first F-16 arrived in November. Conversion was completed during March 1993. The Block 30 F-16C/Ds are marked with a 'WI' tailcode, and a red fin-stripe bearing the name 'Madison' in white, representing the local city.

The 'Raggidieass Militia' have swapped A-10s for F-16C/Ds, which are operated in an air interdiction/close support role.

132nd Fighter Wing, Iowa ANG

Des Moines MAP, Iowa

124th Fighter Squadron

Both Iowa Guard units transitioned from the much-loved A-7 to the 'Electric Jet'. The unit's 'Hawkeyes' nickname is carried at the base of the fin and the local city name acrossd the top.

Like its sister squadron in the Iowa ANG, the 124th TFS was a stalwart of the fighter-bomber force, flying F-100s and later A-7s. It was in the last group of Corsair units to re-equip with the F-16, and bypassed all earlier variants of the Fighting Falcon to convert directly on to the current F-16C/D Block 42 machines. The conversion was accomplished during 1993. The commander's aircraft originally received full-colour markings with a blue/white/red vertical flash on the rudder. A tricolour flag is now carried on all the aircraft, with the state name superimposed. On account of the marking the unit is often referred to as the 'French air force'.

138th Fighter Group, Oklahoma ANG

Tulsa IAP, Oklahoma 'OK'

125th Fighter Squadron

From its Tulsa base the 125th TFS flew the A-7D Corsair II on close air support duties from 1978 to 1993. For the last six years the squadron was one of only three equipped with LANA-capable A-7s. On conversion to the 'Viper' in 1993, the 125th FS acquired F-16C/D Block 42 aircraft, powered by the Pratt & Whitney F100-PW-220 and compatible with the LANTIRN system. The markings consist of a small script 'Tulsa' in white at the top of the fin and the 'OK' tailcode with a white shadow on the fin. The aircraft serial number is also presented with a white shadow.

In 1973 the 125th converted from Globemasters to F-100s, and has been lucky enough to be a fighter-bomber unit ever since.

140th Fighter Wing, Colorado ANG

Buckley ANGB, Colorado 'CO'

120th Fighter Squadron

The 120th TFS rose to fame in the mid-1950s when the squadron formed the Guard's official display team (the 'Minute Men') flying F-80s. After a spell as an air defence outfit, the 120th became a fighter-bomber unit and flew on active duty with F-100s over Vietnam. A-7 Corsairs followed from 1974 until 1992, during which time the squadron won the coveted Gunsmoke trophy in 1981. The first of the unit's F-16C/D Block 30s arrived on 28 August 1991, while the last A-7 left Buckley on 5 February 1992. A successful conversion to the Fighting Falcon was crowned in 1993 by the naming of Major Gregory Brewer as the Gunsmoke 'Top Gun'. 120th FS aircraft are marked with a mountain scene and a script 'Colorado' at the top of the fin, a drop-shadow 'CO' tailcode and the nickname 'Mile High Militia' along the base of the fin. The commander's aircraft also has the white and blue state colours on the fin, and the golden 'C' from the state flag. The 120th was one of four Guard squadrons to staff a detachment to Incirlik for Provide Comfort, and in 1995 six aircraft deployed to Australia for Exercise Down Under.

The Colorado ANG flies from Buckley, close to the state capitol Denver. Honours include a combat detachment to Turkey.

131

142nd Fighter Group, Oregon ANG

Kingsley Field, Klamath Falls, Oregon

114th Fighter Squadron

Part of the 142nd FG, which flies F-15s from Portland, the 114th was established at Klamath Falls on 1 February 1984 as the 114th Tactical Fighter Training Squadron, with the task of training air defence crews for the Air National Guard. Initial equipment was the F-4C, and the 114th was the last USAF unit to fly this variant. The first F-16A/B Block 15 arrived to begin replacement of the F-4 in late 1988, a process completed the following year. On 1 March 1989 the 114th was the first ANG unit to receive an example (82-1041) of the ADF version, and by the fall the whole unit was operating the upgraded variant. On 15 March 1992 the unit was redesignated as a Fighter Squadron. Aircraft are marked with a large bald eagle with outstretched talons, with the word 'Oregon' at the top of the fin.

Previously designated a Tactical Fighter Training Squadron, the 114th FS continues to train F-16ADF pilots.

144th Fighter Wing, California ANG

Fresno Air Terminal, California

194th Fighter Squadron

From its designation as a Fighter Interceptor Squadron in July 1955, the 194th flew a succession of air defence aircraft (F-86A, F-86L, F-102A, F-106A and F-4D). During 1989 the unit converted to the F-16A/B Block 15 (later upgraded to ADF standard), the process being completed in October. With the state name on the base of the fin,

the aircraft carried a striking eagle motif. Operating from Fresno Air Terminal in the Central Valley, the 194th also maintained an alert detachment at George AFB, although on 15 June 1992 this was relocated to March AFB on the closure of George. In mid-1994 the squadron began conversion to the F-16C/D Block 25 as part of the partical dismantling of the air defence network

California's 144th FIW flew the ADF variant from Fresno, forming the southwestern link in the air defence chain. Today the unit has a multi-role tasking with F-16C/Ds.

147th Fighter Group, Texas ANG

Ellington Field, Houston, Texas

111th Fighter Squadron

The 111th 'Ace in the Hole' squadron moved to its present base at Ellington Field (then an AFB) in 1956, and operated as an interceptor unit with, successively, F-80Cs, F-86Ds, F-86Ls, F-102As, F-101B/Fs, F-4Cs and F-4Ds. It also had T-33s while acting as the ANG Jet Instrument School, and undertook the ANG conversion role for the F-101 and F-102. The first F-16A/B Block 15 arrived on 1 December 1989, and during 1990/91 the unit's complement was upgraded to ADF standard. The F-16s wear

the word 'Houston' at the top of the fin, and a stylised presentation of the Texas 'Lone Star' flag at the base. In common with other units, the commander's aircraft is marked in full colour, while regular squadron machines have low-visibility markings. To celebrate the unit's 75th anniversary in 1992, one F-16 received a large flag under the intake, '75 Years' across the fin and the dates '1917-1992' on the ventral fins. In addition to their regular air defence tasking, the 111th FS is heavily involved in the war against drug trafficking. F-16s fly anti-drug patrols along the southern border, and the unit flies the sole UC-26C Metro, fitted with F-16 radar and a FLIR.

The 111th FS flies a secondary anti-drug mission from its Houston base.

148th Fighter Group, Minnesota ANG

Duluth IAP, Minnesota

179th Fighter Squadron

Based since 1948 at Duluth, the 179th's air defence heritage covers operations with the F-51D, F-94A/B, F-94C, F-89J, F-102A and F-101B/F. From 1975 to 1983 the unit flew tactical reconnaissance missions with the RF-4C, before returning to the air defence world with the F-4D. In early 1990 the unit began to receive F-16A/B Block 15 ADFs, the last Phantom flying out in April. The Fighting Falcons carry the city name in a band at the top of the fin and a representation of the Little Dipper constellation. The highlighted North Star signifies the north-facing tasks for the air defence F-16s.

The Duluth squadron flies the ADF F-16, characterised by the bulge at the base of the fin and the AIFF aerial array forward of the cockpit.

149th Fighter Group, Texas ANG

Kelly AFB, Texas 'SA'

182nd Fighter Squadron

In 1956 the 182nd moved to its present base at Kelly AFB, and established as an interceptor unit. The tasking switched to a fighter-bomber role in 1969, and the 182nd flew F-84Fs, F-100Ds and F-4Cs in this role until the mid-1980s. In early 1986 F-16A/B Block 15s began arriving to replace the Phantoms, the squadron completing the

The 182nd FS has operated the Block 15 version of the Fighting Falcon for nearly 10 years. The unit has a true dual-role mission, training for both battlefield air superiority and tactical air-to-ground missions. In the latter the Block 15 aircraft is used as a 'dumb' bomber, having no means of precision attack.

process on 1 July. Since that time the squadron has continued to use this variant on tactical battlefield duties. Aircraft carry a bald eagle motif with state name on the fin-band, and the 'SA' tailcode (for the nearby city of San Antonio).

150th Fighter Group, New Mexico ANG

Kirtland AFB, New Mexico 'NM'

188th Fighter Squadron

Originally an interceptor unit (with F-51D, F-80A/C and F-100A), the 188th became a tactical fighter-bomber unit in 1964 when it adopted F-100Cs. It was on active service in Vietnam during 1968, and converted to A-7Ds in 1973, operating LANA-capable aircraft from 1987. One of the last tranche of Corsair squadrons, the New Mexico squadron converted to the Fighting Falcon during 1992/93, these being F-16C/D Block 40 aircraft with LANTIRN capability. Initially the aircraft carried across the roadrunner badge and Indian motif from the A-7s, together with the state name and the unit's 'Tacos' nickname on the fin. The tailcode 'NM' was carried, and the aircraft featured a yellow fin-band containing a red cross motif. The nickname was carried on the fin/fuselage fairing. A black fin-band with a roadrunner began to be adopted in early 1995.

Equipped with an AIS wingtip pod and an SUU-20 practice bomb carrier, this New Mexico aircraft displays the original yellow tail markings.

The roadrunner is the state bird of New Mexico, and has been reincorporated into the black fin markings now being applied.

156th Fighter Group, Puerto Rico ANG

Muniz ANGB, San Juan, Puerto Rico 'PR'

198th Fighter Squadron

Although it has recently voted in a referendum to remain independent, the island state of Puerto Rico is voluntarily associated with the United States, and its National Guard flying squadron was established in May 1946. During its early existence it was an interceptor unit providing the island's air defence (with F-47N, F-86E, F-86D, F-86H and F-104C), but became a close air support unit in 1975 when it switched to A-7s. Although it was never called to federal active duty, the PR

ANG lost eight A-7s in a separatist terrorist attack. During 1992/93 the squadron exchanged its Corsairs for F-16A/B Block 15 ADFs, these being painted with the state name and flag on the fin, the squadron nickname 'Bucaneros' on the intakes and a pirate figure behind the cockpit. The squadron is organised in three flights, and the state name is edged in either blue, yellow or red. The 'PR' tailcode was worn from the outset, signifying the fact that the 198th FS has operated within the tactical organisation despite the fact it operates ADF aircraft.

The coloured band above and below 'Puerto Rico' is in either yellow (as here), red or blue according to flight assignment.

158th Fighter Group, Vermont ANG

Burlington IAP, Vermont

134th Fighter Squadron

Well known as the 'Green Mountain Boys' (derived from the state name), the Vermont

Vermont's F-16 Block 15s initially flew the air defence mission, and were converted to ADF standard.

ANG flew the interceptor mission from Burlington from 1950 to 1974, when they adopted B-57s for target facilities work. In 1982 the F-4D was adopted for multi-role operations, and in 1986 the 134th returned to air defence work with the delivery of F-16A/B Block 15s. The aircraft began arriving in April, and an official conversion ceremony was held in July. The aircraft were marked with an eagle towing a banner bearing the state name, while the unit's nickname adorned the intakes. In the air defence mission the Vermont ANG were highly active for a CONUS-based unit, performing several intercepts of Soviet reconnaissance aircraft shadowing the eastern seaboard. On 1 April 1988 the 134th FIS Det. 1 was established at Bangor IAP, Maine, and during 1990 the unit upgraded to ADF F-16s. With the rundown of the ANG's air defence force in 1994, the 'Green Mountain Boys' switched to the tactical F-16C/D Block 25. On the new aircraft the markings were revised to feature the state name and nickname at the top and base of the fin respectively, and a representation of an original 'Green Mountain Boy' in the middle. These militiamen had captured Fort Ticonderoga from the British in May 1775.

The ADFs were swapped for F-16C/Ds in 1994, the Vermont Guard unit also adopting a new dual-role mission.

133

162nd Fighter Group, Arizona ANG

Tucson IAP, Arizona 'AZ'

148th Fighter Squadron
152nd Fighter Squadron
195th Fighter Squadron
ANG/AFRes Flight Test Center
F-16A/B International Military School

In the summer of 1969 the 152nd FIS swapped its F-102As for a batch of F-100Cs and was redesignated the 152nd TFTS. It immediately began the task of training ANG crews in the Super Sabre. In 1975 it began to add the A-7D training task, although the F-100 was not retired until 1978. The A-7 RTU function continued until early 1984 when it was transferred to a newly-established squadron, the 195th TFTS, also at Tucson under the 162nd TFG. Although the 152nd adopted the training task on the F-16, it did not begin operations until March

1986. In the interim, the 148th TFTS had been newly-created to begin F-16 RTU work for the ANG, making the 162nd TFG a three-squadron group. The 195th TFTS continued to train A-7 pilots until 26 July 1991, at which point it completed its conversion to the F-16. All ANG tactical F-16 training was handled at Tucson, which offered excellent year-round weather and nearby ranges. In January 1990 the 148th TFTS added the role of training pilots from the Netherlands, for which an extra 14 aircraft were supplied, including 10 from the KLu. This detachment ended in April 1995. The F-16A/B International Military School is at Tucson, and by 1996 all F-16A/B foreign training is to

Part of the 162nd FG activity is the ANG/AFRes Test Center, which operates a handful of F-16s.

be concentrated at the base. Another subordinate unit is the ANG/AFRes Test Center, which conducts operational trials with a number of F-16s, including ADF versions. The three RTU squadrons have used the F-16A/B Block 10 for many years, but in 1995 the first F-16C/D Block 25s were delivered. No individual squadron markings are worn, all aircraft being marked with the sunset state flag and 'Arizona' on the fin. The tailcode appears on only some aircraft (mostly F-16C/Ds).

The 162nd FG has three squadrons for ANG F-16 training. The unit also trained Dutch pilots, the aircraft so employed having been supplied by the KLu, marked with a small Dutch roundel on the side of the intake (above).

169th Fighter Group, South Carolina ANG

McEntire ANGB, South Carolina

157th Fighter Squadron

In 1975 the 157th FIS left behind its F-102s to become a tactical unit flying A-7Ds. These served until 1983, when the unit was given the honour of becoming the first ANG

unit to operate the F-16, its first F-16A/B Block 10 aircraft arriving in July. The 'Swamp Foxes' became very proficient in the 'Viper', taking team honours in the 1989 Gunsmoke competition. Their 'reward' was to be called to Federal Active Duty on 29 December 1990 and dispatched to Al Kharj in Saudi Arabia for Desert Storm, operating

as part of the 4th TFW(P). After flying 1,729 combat sorties, the 157th TFS was returned to state control on 22 July 1991. The 157th (which became an FS in March 1992), had operated the same Block 10 aircraft for over 10 years, but in 1994 the unit received much-needed new equipment in the form of the current Fighting Falcon variant: the F-16C/D Block 52. Like their predecessors, the new aircraft carry the state name in a

blue band across the fin and the 'Swamp Fox' motif behind the cockpit. The 'SC' tailcode was worn only in the early days of F-16 operations.

The fins of these South Carolina aircraft proclaim them to be 'World Champions – Gunsmoke 89'. The unit fought in Desert Storm.

Above: Current equipment for the 157th FS is the Block 52 F-16C/D, powered by the Pratt & Whitney F100-PW-229 IPE.

174th Fighter Wing, New York ANG

Hancock Field, Syracuse, New York 'NY'

138th Fighter Squadron

Established as a fighter-bomber unit with P-47Ds and then F-84Bs, the 138th 'Boys from Syracuse' then had a brief period as an interceptor unit before returning to the tactical fold in 1958, flying F-86Hs, A-37Bs and A-10As. In November 1988 F-16A/B Block 10s began arriving at Syracuse, the conversion being completed on 31 December. The 138th's aircraft were unique insofar as they were the only operational F-16s wired to carry the GPU-5/A Pave Claw gun pod. With this weapon the unit became a pioneer of the 'Fastass CAS' mission, and a year after getting the aircraft was called to active duty for the Gulf

War. The F-16s departed Hancock Field on 2 January 1991, and were in place at Al Kharj with the 4th TFW(P) to begin offensive operations on Day One of Desert Storm. The squadron flew 1,050 combat sorties, although the Pave Claw, which had proved disappointing in service, was believed to have been carried on only one day. A mixture of bombs and AGM-65 missiles was carried for most of the war. A return to state control on 30 June 1991 was followed by conversion in 1992/93 to the F-16C/D Block 40. Tail markings on both variants of F-16 comprised a blue-white checkerboard fin-stripe, the 'NY' tailcode and the unit's nickname along the fin/fuselage fairing. A falcon badge behind the cockpit on early aircraft was subsequently replaced by a cobra.

Having previously operated F-16A/Bs in the CAS role, the 'Boys from Syracuse' now fly the F-16C/D.

177th Fighter Group, New Jersey ANG

Atlantic City IAP, New Jersey

119th Fighter Squadron

After a considerable time as a tactical fighter-bomber unit, the 119th TFS swapped its F-105Bs for F-106As in 1972. It flew these until 1 August 1988, becoming the final operational user of the Delta Dart. Replacing them were F-16A/B Block 15s, which went on alert on 1 November. The 119th undertook several intercepts of Soviet aircraft during its interceptor career, and in 1991/92 the unit's aircraft were upgraded to ADF standard. The aircraft were marked with a large fin-flash (red on the commander's aircraft) with the state name across the top of the fin. Midway up the fin was the unit's 'Jersey Devil' badge, although this was later removed. In the light of the reduced air defence requirements,

the 199th transitioned to the F-16C/D Block 25 in the winter of 1994/95.

After 23 years as an air defence unit, the 119th FS has recently exchanged these ADF aircraft for tactically-roled F-16C/Ds.

178th Fighter Group, Ohio ANG

Springfield-Beckley MAP, Ohio 'OH'

162nd Fighter Squadron

From 1962 the 162nd has been one of the stalwarts of the ANG's fighter attack force, flying F-84Fs, F-100Ds and A-7s. It was the final user of the A-7 in USAF service, operations with the Corsair ending in 1993. These were replaced by F-16C/D Block 30s. Tail markings consist of a red/yellow checkerboard fin-stripe, 'OH' tailcode (shared with the 112th FS) and a representation of the unit's black panther badge. On the starboard intake side is the squadron badge, while on the port side of the intake is a winged gauntlet holding a sword and the legend 'Yoxford Boys'. This commemorates the unit's period of service flying P-51Ds from Leiston in England in 1944/45.

This 162nd FS aircraft wears a kill marking for a victory scored over an Iraqi MiG-29 during Provide Comfort operations on 16 January 1993, while serving with the 52nd FW.

180th Fighter Group, Ohio ANG

Toledo Express AP, Ohio 'OH'

112th Fighter Squadron

Like its sister Ohio squadron, the 112th transferred to the Fighting Falcon from the A-7D, having flown the Corsair in action over Grenada. Conversion to the F-16C/D Block 25 for battlefield duties began with the first receipt on 29 February 1992 and was completed early the following year. In addition to the 'OH' tailcode, the F-16s carry a green fin-band with the word 'Toledo' and the state outline superimposed, with the unit's 'Stingers' nickname and yellowjacket badge on the fin/fuselage fairing.

Like many ANG units, the Toledo 'Stingers' operate from a base complex attached to a civilian airport.

181st Fighter Group, Indiana ANG

Hulman Field, Terre Haute, Indiana 'TH'

113th Fighter Squadron

Based at Hulman Field, the 113th TFS clung on to its 'HF'-coded F-4Es until 1991, when it converted to the Fighting Falcon. The first F-16 arrived in April, and the last Phantom left in October. The squadron had been a tactical fighter-attack unit since 1958, and initially used the F-16C/D Block 25 in this role. The tailcode changed from 'HF' to 'TH'

Terre Haute is situated about 70 miles from the Indianapolis racetrack where the famous '500' race is held each Memorial Day weekend. The unit has adopted this theme for its markings and nickname.

(for Terre Haute), and the unit applies a fin-band consisting of an eagle's head, the state name and a checkered flag. The latter, and the unit's 'Racers' nickname applied on the base of the fin, associates the unit with the nearby Indianapolis racetrack. In 1995 the squadron began a minor upgrade to the Block 30 variant.

182nd Fighter Group, Illinois ANG (not current)

Greater Peoria AP, Illinois 'IL'

169th Fighter Squadron (not current)

In 1969 the 169th became a specialist air control unit, flying the Cessna U-3, O-2 and OA-37. On 14 March 1992 the unit began its return to the fighter-attack world with the delivery of its first Fighting Falcon, the last OA-37B leaving on 3 June. The aircraft delivered were F-16A/B Block 15 ADFs but, like those of the Puerto Rico ANG, the 169th FS machines were not dedicated to the air defence mission, having instead a dual-role battlefield role. Consequently they wore the TAC-style 'IL' tailcodes from the start. Further markings consist of a fin-band with the word 'Peoria' and the ANG 'Minutemen' logo. The commander's aircraft has full-colour markings including a flaming spear. In late 1994 the unit retired its F-16s in preparation for retasking as a C-130 transport unit.

Despite operating the ADF F-16, the 169th FS was also tasked with air-to-ground missions.

183rd Fighter Group, Illinois ANG

Capital MAP, Springfield, Illinois 'SI'

170th Fighter Squadron

From its Springfield base, the 170th has conducted fighter operations since 1948, mostly in the tactical arena. It was the last unit to fly the F-84F before adopting the F-4C and F-4D. The switch to F-16s occurred in 1989, conversion to the Block 15 variant being completed on 1 October. Markings consist of the 'SI' tailcode (for Springfield, Illinois) and multi-hued fin-stripe with the state name. From November 1994 the squadron received F-16C/D Block 30s.

These smart F-16As of the 170th FS were Block 15s assigned a dual-role battlefield mission. The markings are presented in toned-down form here, although the full-colour version has a red, white and blue fin-stripe and legend. These aircraft also carry the Illinois state silhouette on the fins of the drill Sidewinder rounds, superimposed with '170'. Here the aircraft are launching for a practice bombing sortie, carrying BDU-33 25-lb 'blue bombs' from triple ejector racks. Today the unit has the same role, but now has the upgraded F-16C/D.

184th Fighter Group, Kansas ANG (not current)

McConnell AFB, Kansas

127th Fighter Squadron (not current)
161st Fighter Squadron (not current)
177th Fighter Squadron (not current)

The 184th TFG at McConnell AFB was chosen in 1987 to be the second training unit (after Tucson) for the ANG F-16 force. The group's 127th TFTS had been an F-105 RTU in the 1970s and had performed the training task on the F-4D. A second F-4D squadron had been formed in 1984 as the 177th TFTS. While both these units remained active on the Phantom, the first F-16A/B Block 10 was delivered to the newly-created 161st TFTS, which was activated on the type on 1 July 1987. The 127th transferred to the F-16A/B in the fall, while it was not until March 1990 that the

177th TFTS phased out the F-4D in favour of the F-16. The three squadrons operated aircraft on a pooled basis, and no individual squadron marks were applied. All wore a 'Kansas' fin-stripe and the 'Jayhawks' nickname. During the winter and spring of 1990/91 the McConnell training group changed over to the F-16C/D Block 25 to

Like the Arizona training unit, the Kansas group had three squadrons but all the aircraft wore similar markings. This is an F-16C.

When first established, the 184th TFTG operated the F-16A/B. A large proportion of the fleet were two-seaters.

reflect the large number of second-generation F-16s entering ANG service. The unit continued on training duties until 1994, when it was disbanded in the light of reduced pilot requirements. The 184th became a Bomb Group, operating B-1Bs, while the F-16C/D training mission was integrated with the F-16A/B effort at Tucson.

185th Fighter Group, Iowa ANG

Gateway AP, Sioux City, Iowa

174th Fighter Squadron

The 174th has been a ground attack unit from 1961, flying the F-100C, F-100D and A-7D (including LANA aircraft from 1987). On 19 December 1991 the unit took delivery of its first Fighting Falcon, despatching its last A-7 to the boneyard on 20 February 1992. The aircraft are F-16C/D Block 30s, and were given a striking fin marking consisting of a black bat, the state name and 'Sioux City'. The nickname 'The Bats' is carried behind the cockpit. It is the unit's base at Sioux ('Gotham') City which provides the bat theme. The 174th was one of four ANG squadrons to provide a detachment to Turkey for Provide Comfort in 1994.

Most of the 'Bats' aircraft from Sioux ('Gotham') City wear individual nose art.

187th Fighter Group, Alabama ANG

Dannelly Field, Montgomery, Alabama 'AL'

160th Fighter Squadron

From 1950/51 until 1983 both squadrons of the Alabama ANG were employed on tactical reconnaissance duties. While the 106th remained on RF-4Cs until the 1990s, the 160th converted to a fighter-bomber role with F-4Ds. These were replaced by F-16A/B Block 10s in 1988, the unit

The 160th FS has recently completed the conversion from the Block 10 F-16A/B to the F-16C/D Block 30 (illustrated).

completing its conversion on 1 October. In addition to the 'AL' tailcode, the aircraft wore 'Montgomery' in a fin-band and a cobra (from the squadron badge) behind the cockpit. These long-serving F-16s were replaced by F-16C/D Block 30 aircraft in 1993.

188th Fighter Group, Arkansas ANG

Fort Smith MAP, Arkansas 'FS'

184th Fighter Squadron

Nicknamed the 'Flying Razorbacks' (in recognition of the wild hogs of the Ozark mountains), the 184th TFS converted to the F-16A/B Block 15 from the F-4C in early 1989, its official conversion date being 1 April. The aircraft are operated in the dual-role mission. Standard squadron aircraft wear their markings in dark grey, consisting of a fin-stripe with the state name and the nickname on the fin/fuselage fairing. The 'FS' tailcode stands for Fort Smith. The commander's aircraft has full-colour markings, also adding a shooting star containing the squadron's razorback insignia.

In the centre of the fin-star is the 184th FS razorback hog badge, which is also repeated on the intake side. This machine has its F100 engine removed for maintenance.

191st Fighter Group, Michigan ANG (not current)

Selfridge ANGB, Michigan

171st Fighter Squadron (not current)

After a spell as a reconnaissance outfit, the 171st 'Michigan Wolves' returned to its interceptor mission in 1972, flying F-106As, F-4Cs and F-4Ds before adopting the Fighting Falcon. Conversion to the Block 15 version was completed on 1 July 1990, and the aircraft were subsequently upgraded to ADF standard. In terms of markings the squadron aircraft had a double fin-flash in dark grey with a checkerboard pattern on the rudder and the state name on the top of the fin. The checks were repeated on the fillet between the intake and fuselage. The commander's aircraft featured these markings in yellow and black. In 1994 the 171st FS was the first of the air defence squadrons to be deactivated.

Marked in full-colour, this F-16B ADF was the 171st commander's aircraft. The unit is now a Hercules transport squadron.

192nd Fighter Group, Virginia ANG

Byrd IAP, Richmond, Virginia 'VA'

149th Fighter Squadron

In 1982 the 149th converted from F-105s to A-7s, and flew these from Richmond until the last left on 1 October 1991. Hastening their departure was the delivery from 25 June 1991 of F-16C/D Block 30s with which the 149th continued the fighter-bomber mission after completing conversion on 1 June 1992. In 1994 the squadron helped man the ANG combat deployment for Operation Provide Comfort, protecting the Kurds in the northern Iraq 'No-Fly Zone'. In 1995 the Virginia ANG was chosen to test a Lockheed Martin multi-sensor reconnaissance pod on an F-16 which the company is offering to the USAF to fill a tactical reconnaissance requirement caused by the virtual retirement of the RF-4C. Since entering service, the Virginia F-16s have worn a 'VA' tailcode and the state name in a smart fin-stripe.

The Virginia ANG conducted trials in 1995 with a Lockheed reconnaissance pod under an F-16D.

Armed with an AGM-65 Maverick training round, this 149th FS F-16C prepares for a close air support training mission.

Spain

Isolated under Franco, Spanish links with the rest of Europe have multiplied since the restoration of democracy. Since November 1988, when Spain joined the Western European Union, its armed forces have been loosely committed to the mutual defence of the Benelux countries, France, Germany, Portugal and the UK. This link has led to the ongoing Spanish participation in UN/WEU operations over Bosnia. Spain is notionally a member of the NATO military structure, though its forces are not formally assigned. NATO membership has given Spain access to assistance from other members, and to modern technology and weapons systems which have allowed a rapid modernisation of the armed forces.

Ejército del Aire (Spanish air force)

Spain's first contact with military aviation goes back to December 1896, when an Army balloon unit, the Servicio Militar de Aerostación (Military Aerostation Service) was activated. Not until 1911 was a Servicio Militar de Aeronáutica (Military Aviation Service) formed, with the inauguration in March of an Aviation School at the now historic air base of Cuatro Vientos near Madrid. By 1918 the Aeronáutica Militar had five operational squadrons and grew steadily during the following years, receiving its baptism of fire in Morocco. By 1929 some 400 aircraft of all types were in service, most of them obsolete types.

In 1931 Monarchy gave way to Republic in Spain and a limited modernisation was envisaged, including the possibility of building the Hawker Fury fighter under licence; with the withdrawal of obsolete types, strength was trimmed to 287 machines. Spain had suffered heavily from endemic political instability since the late 1890s, and the several republican governments that came after 1931 could not stabilise the country's internal situation. The inability of the elected government to reconcile the demands of the disaffected right, the workers and regional separatists against the background of a depressed economy and growing unrest led to the 18 July 1936 coup by part of the army which hoped to restore order and unity. This involved primarily the forces in Morocco, commanded by General Franco, supported by rightist politicians and the Catholic church.

The Aeronáutica Militar came to an end, the title being adopted by Spanish Nationalist aviation units which merged with those navy units that supported the uprising. Loyal government aviation units came under the new Fuerzas Aéreas de la República Española (FARE). The Nationalist side received generous help from Italy who sent a contingent known as the Aviación

Legionaria, and Germany who sent the Legion Condor. The Republicans received aircraft and equipment from predominantly Russian sources. The Civil War ended on 29 March 1939 with the victory of Nationalist forces, and on 9 November 1939 a new air arm was formed as an independent service: the Ejército del Aire. Since then the Spanish air force has experienced neutrality during World War II, isolation from 1945 to 1953, entering the jet era after signing a defence agreement with the United States which saw massive infusions of F-86 Sabres, T-33, T-6s, etc. Consolidation during the 1970s and 1980s resulted in the mature air arm of today.

The modern air force

Perhaps the most vital step has been the acquisition of McDonnell Douglas F/A-18 Hornets, which has pulled the entire Ejército del Aire into the high-tech era. This impacyed on everything that surrounds such a sophisticated weapons system: component production, repair and overhaul, simulators, software, etc. These spin-offs are being exploited and the country's capabilities are expanding, with Spain becoming a partner in the EF2000 and FLA programmes.

Since 1982 Spain has been a NATO member. Although it is not formally or politically integrated within NATO's military structure, collaboration by the Ejército del Aire with allied air arms has grown dramatically during recent years due to the withdrawal of Canadian and US air assets from Europe. Proof of that is the F-18/KC-130 detachment at Aviano air base, northern Italy, for use in Deny Flight operations over Bosnia. Other Spanish aircraft, mainly transports, have participated in multinational/UN humanitarian operations anywhere in the world. This has been one of the peace dividends of the end of the Cold War, and the Ejército del Aire is quickly adapting

to its new roles with combat-ready, highly mobile forces. According to the Ministry of Defence Joint Strategic Plan, the main threat to Spain's security in the future may come from North Africa. Nevertheless, co-operation exists with air arms in this area by means of the yearly Atlas air exercises with the Royal Moroccan air force and regularsquadron or pilot exchanges.

Regional command structure

Currently, the Ejército del Aire has more than 500 aircraft and a professional strength of some 20,000 personnel. Under its Cuartel General (Headquarters) in Madrid it is divided, after the 1989/90 reorganisation, into regional commands: Mando Aéreo del Centro (**MACEN**) – Central Air Command with HQ at Madrid; Mando Aérea de Levante (**MALEV**) – Eastern Air Command with HQ at Zaragoza; Mando Aéreo del Estrecho (**MAEST**) – Southern Air Command with HQ at Seville; and Mando Aéreo de Canarias (**MACAN**) – Canaries Air Command with HQ at Las Palmas de Gran Canaria. Most of the flying and ground units are attached to these commands, which are complemented by three special-purpose commands: Mando Aéreo de Personal (**MAPER**) – Personnel Air Command with HQ at Madrid; Mando Aéreo Logéstico (**MALOG**) – Logistics Air Command with HQ at Madrid; and Mando Operativo Aéreo (**MOA**) – Operational Air Command with HQ at Torrejón air base.

MAPER exercises functional control over the Ejército del Aire training assets in five air bases, including the Academia General del Aire at San Javier; Ala 78 at Armilla, Granada; Ala 23 at Talavera, Badajoz; Grupo 74 at Matacán, Salamanca; and Grupo 42 at Getafe, Madrid. It also provides resources and training systems.

MALOG is the organisation responsible for logistical support and infrastructure. Under its charge are three Air Rework Facilities which perform overhaul, repair and modification to the air force fleet, engines, electronics, etc. These facilities are Maestranza Aérea de Madrid at Cuatro Vientos air base; Maestranza Aérea de Albacete; and Maestranza Aérea de Sevilla, supported by civilian companies CASA and AISA.

MOA is a new organisation, which in time of peace is in charge of surveying Spanish airspace and of overseeing the training of combat units to reach the desired readiness level. Also, it is the organiser of the bi-monthly air defence exercises (DAPEX for the mainland and DACEX for the Canary Islands).

In time of crisis MOA takes operational control of front-line assets. In peacetime it is responsible for eight radar stations – three on the islands (Balearics and Canaries) and five on the mainland – to which another three will be added soon, two on the mainland and the third on Isla del Hierro, Canaries. These stations are known as Escuadrones de Vigilancia Aérea (Air Surveillance Squadrons) and the data provided, besides being used by local controllers at the sites, are sent to a SOC (Sector Operations Centre) and a COC (Combat Operations Centre) which are the responsibility of the Grupo Central de Mando y Control (GRUCEMA) (Command & Control Group) established at Torrejón air base in hardened underground installations. The early warning system, 'Combat Grande', is operated automatically and is steadily being improved. It is now

Grupo 12 was Spain's first Hornet unit, and the first F/A-18s (or EF/A-18s) arrived in 1989 to replace veteran F-4Cs (known as C.12s in Spanish air force service). Grupo 12 has three component escuadrones: Nos 121, 122 and 124.

Above: The Torrejón Hornets are part of a virtual 'superwing', as they operate in conjunction with Grupo 12's RF-4Cs and the 707 tanker/transports of co-located Grupo 45.

Below: The eight RF-4Cs of Escuadron 123 were formerly operated by the 123rd TRW, Kentucky ANG. These Phantoms were delivered to Ala 12 in 1989.

Grupo 12's Hornets are shared between its three escuadrones, and all aircraft wear the Grupo's black panther badge. Esc 121 (callsign 'Poker') and Esc 122 (callsign 'Tenis') are the two former F-4 operators.

The Dornier Do 27 STOL transport (U.9) serves with several Ejército del Aire units as a general transport and liaison aircraft. Those attached to Grupo 12 wear a grey scheme that reflects the combat role of their parent unit.

Spain's first 'recce' Phantoms were four RF-4Cs (CR.12) delivered in 1978, as attrition replacements for its F-4Cs. The current aircraft arrived in January 1989, for Esc 123 (callsign 'Titan'), and have all been upgraded through the addition of AN/APQ-172 TF radar, new ECM and INS systems, EO sensors, datalinks and Israeli-style refuelling probes.

Grupo 43's winter home is Torrejón, but the unit's CL-215 fire bombers deploy all around the country during the summer months, to combat forest fires. Of the 30 aircraft originally acquired, from 1973 onwards, seven have been lost in the course of their hazardous and demanding job.

In 1989 Canadair introduced a turboprop upgrade for the CL-215, the CL-215T, which adds PW123AF engines and several aerodynamic refinements including the sizeable winglets so apparent here. Spain ordered 15 upgrade kits for delivery by 1993 – the only CL-215 customer to have done so thus far.

running under the SIMCA (Integrated Command & Control) programme, with hardened communications and new 3-D radars with ECCM capability are being incorporated.

MOA is also responsible for the QRA reaction force in the mainland, composed of one pair of alert fighters in each of Ala 14, Grupo 15 and Ala 12. (In the Canaries this is the responsibility of the autonomous Early Warning Group, which also has fully automatic facilities and is located at Gando air base, maintaining one Mirage F1 from 462 Escuadrón at 24-hour readiness.) MOA has one fighter squadron directly assigned, 141 Escuadrón of Ala 14, whose Mirage F1s have Elint capability with Thomson-CSF Syrel pods. Finally, the Ejército del Aire's premiere electronic warfare unit, Torrejón-based 408 Escuadrón equipped with Aviocars, Falcon 20s and soon an IAI-modified Boeing 707, is assigned to MOA.

At unit level the Ejército del Aire is organised into Alas (wings). Typically, each wing has three groups – Grupo de Fuerzas Aéreas (Operations Group) in charge of flying operations, Grupo de Material (Material Group) in charge of maintenance and supplies, and Grupo de Apoyo (Support Group) which supports the base infrastructure, vehicles, security, etc. Although pilots are assigned to a particular squadron, aircraft have had their individual squadron numbers changed to reflect that of the parent wing or group. As an example, Ala 12 at Torrejón has three operations groups, Grupo 12 with F/A-18A/B+s and RF-4Cs, Grupo 43 with CL-215Ts, and Grupo 45 with Falcon 20/50/900s and Boeing 707s, plus a single independent squadron, 408 Escuadrón with Aviocars, Falcon 20s and a Boeing 707. Ala 31 at Zaragoza is another example of 'big wing' operations, with Grupo 15 operating Hornets and Grupo 31 with C-/KC-130Hs. All aircraft at a given base belong to the Maintenance Squadron of the Material Group, which then provides available ones to the flying units.

Fast jet force

Undoubtedly, the Ejército del Aire's most important asset is the McDonnell Douglas F/A-18 Hornet. During the period 1986-90, a total of 72 was received, comprising 60 single-seat F/A-18As and 12 two-seat F/A-18Bs. In nine years of flying three aircraft have been lost in accidents, two from Grupo 15 and one from Ala 12. The 69 surviving Hornets are shared by four front-line squadrons and one OTU squadron.

Beginning in late 1993 the Ejército del Aire undertook a programme to equip its fighter unit at Morón, Grupo 21, which recently had phased out of service its tired CASA/Northrop F/RF-5As and temporarily re-equipped with unarmed C.101EB Aviojets for the tactical training role. This programme, known as CX, envisaged acquiring 18 to 36 secondhand fighters from American sources, either F-16A Fighting Falcons or F/A-18A Hornets; the decision fell on the latter, and from mid-1996 24 Hornets will be delivered, with options on six more. Once in Grupo 21 service the aircraft will be progressively retrofitted. They will arrive with 3,000 flight hours on their airframes, so are good for at least another 3,000 hours or 15 years of operations. Being from early US Navy blocks they have the XN-5 mission computer and the OFP-87X software tape, but their weapons, ECM suite and FLIR/LTD will be the same as (or will be modified/updated to

bring them to the same standards as) examples in the original Spanish fleet.

The first unit to receive the Hornet in Spain was Ala 15 at Zaragoza air base, while the first four F/A-18Bs arrived in flight from St Louis on 10 July 1986. Gradually the number of aircraft being operated by 151 Escuadrón grew. 152 Escuadrón was activated in July 1989, and a month later Ala 15 became Grupo 15 of Ala 31, which also comprised Grupo 31 with Hercules at Zaragoza. Meanwhile, both Escuadrones 121 and 122 of Ala 12, based at Torrejón and previously equipped with F-4C Phantoms, started to convert to the F/A-18 and sent pilots and ground crews to Zaragoza on rotational basis. On 28 March 1989 the first seven F/A-18As and five F/A-18Bs arrived at Torrejón from Zaragoza, with deliveries completed about eighteen months later. Grupo 15 QRA was established in March 1988, while that of Ala 12 got into action from late 1989.

Capable Hornets

The F/A-18's main asset is performing both the air-to-air and air-to-ground roles. Each pilots flies 160 to 180 hours per year, and both units dedicate 60 per cent of their flying time to air-to-air operations and 40 per cent to air-to-ground. In the former role, the aircraft have recently acquired AIM-120A AMRAAM capability to complement the AIM-9M Sidewinders and AIM-7Ms. For air-to-ground operations, the Hornet is equipped with cluster munitions and the Paveway II series of LGBs (GBU-10/16), in addition to low-drag and retarded 'dumb' bombs. Illumination is provided by the AN/AAS-38A NITE Hawk system which combines the functions of a FLIR and a laser target designator and rangefinder. Hornet air-to-surface missiles include the IR-imaging AGM-65G Maverick for use against ground or naval targets, the AGM-84 Harpoon for anti-shipping operations and the AGM-88 HARM for 'Wild Weasel' missions. There is also the M61A1 20-mm available for both air-to-air and air-to ground work.

The 72 Spanish Hornets built at St Louis belonged to Lots 8 to 12, or Block 17 to 31, so it was necessary to bring the whole fleet to a common standard of the later machines, known as F/A-18A+/B+, which are very close to the USN's F/A-18C/Ds. From 15 September 1992 to mid-1995, all the aircraft received a retrofit by McDonnell Douglas and CASA engineers at Grupo 15's Maintenance Squadron hangars at Zaragoza; this basically consisted of rewiring the aeroplane and installing several modifications of software and hardware, including a new mission computer, and fitting new stick tops compatible with the AIM-120A. Since arriving in Spain, the Hornet's software has been of US Navy origin, from the early OFP-84A to the last OFP-89C, but by early 1996 all Hornets will be loaded with the Spanish-designed OFP-94E, which is better suited to Spanish weapons. The F/A-18 electronic warfare set consists of the ALR-67 radar warning sets, ALE-39 flare/chaff dispensers and the ALQ-126B internal jammer, although the more advanced ALQ-162 has been retrofitted to about half of the fleet. During 1999/2002 the Hornets will undergo a Mid-Life Upgrade (MLU) which will make them compatible with new weapons and possibly some elements of the new APG-73, and will extend their lives beyond the year 2010.

Grupo 15, which follows the traditions of Ala

de Caza 2 which was equipped with Sabres in the 1950s and 1960s, has an established strength of 28 F/A-18A+s and two F/A-18B+s. They are shared by both its squadrons, each of which has different taskings. 151 Escuadrón is primarily charged with all-weather fighter-interceptor missions (AWX), including offensive and defensive counter-air (OCA/DCA), and secondarily performs fighter-bomber duties. 151 is assigned to the new NATO Rapid Reaction Forces (Air), and from 1 December 1994 to 31 March 1995 – together with its sister 152 Escuadrón – participated in Operation Deny Flight with eight Hornets deployed to Aviano in Operación Icaro. The unit performed daily air-to-air and air-to-ground CAPs over Bosnia while in permanent contact with FAC controllers, both Spanish and other nationalities attached to UNPROFOR troops in the field. The main task of 152 Escuadrón is SEAD (Suppression of Enemy Air Defences) for which it intensively trains in HARM and LGB operations. Secondarily, the squadron has the same taskings as 151. Both units are tasked in rotation for the QRA commitment.

Torrejón wing

At Torrejón, Ala 12 has 29 F/A-18A+s and 10 F/A-18B+s shared by three squadrons. 121 Escuadrón's main task is TASMO (tactical air support for maritime operations), for which the Maverick, Harpoon and LGB are the main attack weapons, usually working in unison with the P-3 Orions of Grupo 22. TASMO also includes offensive counter-air (OCA) operations over the sea. Secondarily, the unit performs AWX, offensive air support (OAS) and air interdiction (AI). 122 Escuadrón is tasked primarily with AWX, and secondarily with the missions also assigned to 121. This is the second Spanish Hornet squadron assigned to the NATO's Rapid Reaction Forces. 124 Escuadrón, activated on 1 November 1992, is known as the Unidad de Transformación Temporal (Operational Conversion Unit), its instructors being experienced pilots from the four first-line squadrons who are seconded on four-month tours. Assigned most of the two-stick Hornets, 124 provides the basic Hornet course to pilots newly posted to any of the four squadrons: this lasts nine months and includes classroom, simulator missions (there is a simulator at Torrejón and another at Zaragoza) and 61 sorties. After completion the pilot is declared LCR (Limited Combat Ready) and continues operational training in his squadron until reaching the maximum level, CR3, some 18 months later. During July 1994, Euadrones 121 and 122, with three pilots of 152 attached, participated for the first time in a Red Flag exercise at the USAF's Nellis AFB. One of Ala 12's Hornets, configured for SEAD missions, is on a permanent loan to the Torrejón-based Air Weapons and Testing Centre (CLAEX).

123 Escuadrón, equipped with eight RF-4C Phantoms, is tasked with tactical reconnaissance operations. These aircraft were delivered from ANG stocks early in 1989 (previously, from 1972 Ala 12 had operated 40 F-4Cs and four RF-4Cs), having been refitted with smokeless engines, secure radios, VOR/ILS and new RWRs. The camera equipment consisted of KS-87 cameras with an advanced IR detecting set. Soon after arrival they received an overall grey scheme similar to that of the F/A-18s. In order to remain as

Spain currently has three tanker/transport Boeing 707s (T.17) in service with Grupo 45. This is the youngest example of the trio, a 707-368C that was delivered to Saudia in 1977 and acquired by the air force in 1990. Fitted with Sargent-Fletcher wingtip pods, the 707s have been used to support EF/A-18 deployments to Reg Flag and elsewhere.

Two of the Spanish T.17s are combi 707-300Cs, while one is an all-passenger 707-300B. The latter is the VVIP transport which replaced a DC-8-52. Grupo 45 will soon acquire an Elint-configured 707, designated TM.17.

Regular ministerial transport duties are undertaken by a fleet of Dassault Falcons. Falcon 50s are flown by Esc 452, while the larger Falcon 900s are on charge with Esc 451.

Esc 408 is a dedicated EW training unit based at Torrejón, alongside Ala 35. Two specially modified Aviocars (TM.12Ds) were the squadron's first equipment. These were followed by a pair of Falcon 20s (TM.11), and will soon be joined by the new Elint Boeing 707.

Large numbers of the Airtech CN.235 (T.19) are in air force service, and Grupo 35 has two Escuadrones of T.19A/B transports, operating from Getafe. Despite the 'Airtech' moniker, only CASA-built aircraft are in use.

The restrained grey and white scheme of this Grupo 35 T.19 marks it as a VIP-configured T.19A. The tactical transport T.19Bs are all camouflaged.

In recent years Ala 37's Aviocars have seen much service abroad under the banner of the United Nations. White-painted T.17s, such as this aircraft, have been deployed to Namibia and, even more recently, Aviocars have gone to Vicenza, in Italy, with UNPROFOR.

This CASA C.212-AV2-270 (T.12C) is one of the small number of VIP-configured Aviocars allocated to Ala 37, at Villanubla. This wing is the air force's light transport unit, and its aircraft are currently adopting this overall grey scheme.

Ala 48 is charged with a wide range of duties. The Pumas of Escuadrón 402 are tasked with VIP transport and a secondary SAR role. The squadron operates both SA 332B Super Pumas (HT.21) and AS 332M/AS 532 Cougars (HT.21A), the latter type seen here in the foreground. The large fuselage fairing on the HT.21A is an air conditioning unit for the cabin.

viable reconnaissance platforms until 2010 a two-stage upgrade programme was implemented, including the replacement of the APQ-99 terrain-following radar for the APQ-172, a laser ring gyro INS integrated with a new WDNS including a GPS, digital radar altimeter, video recording system and 1553B databus. This permits the use of external mission planning sources, recording data in a tape which is loaded into the aircraft prior to the mission. An Israeli-designed fixed inflight-refuelling probe has been installed, although the USAF refuelling receptacle is also still operational.

Mirage F1

The Ejército del Aire was an early customer of Dassault's second-generation Mirage, the sleek F1. On 1 June 1974, Ala 14 was activated at Los Llanos air base, near Albacete. Some 73 examples were received from April 1975 to April 1983, comprising 45 Mirage F1CEs, 22 F1EEs and six F1BEs. All were subsequently retrofitted with the improved Cyrano IVM radar, as well as the capability to carry the Barax jamming pod. The EEs also have an INS and a fixed inflight-refuelling probe, as well as the capability to operate the Thomson-CSF TMV018 Syrel electronic intelligence pod. Only 30 F1CEs, three BEs and 17 EEs remain in service, having been supplemented between 14 November 1994 and 15 March 1995 by four ex-Armée de l'Air F1Cs and one F1B (in exchange for two CN.235s) to cover attrition. Additionally, six ex-Qatari Mirage F1EDAs and one F1DDA were delivered to 111 Escuadrón of Ala 11 in August/September 1994, with another five EDAs and a single DDAs scheduled to be delivered late in 1997. Mirage F1s equip four squadrons, two at Los Lanos, one in Manises and another in Gando-Las Palmas. The camouflage scheme has been standardised from the original lizard (or blue) scheme to the Celomer PU66 NATO light grey in corrosion-resistant polyurethane, with a false cockpit painted below the original, although the new scheme has yet to be applied fleetwide. Recently, six F1EEs received new RWR sets in place of the standard Thomson-CSF systems: the Indra AN/ALR-300 – which was designed and manufactured in Spain – performs similarly to, and in some instances better than, the ALR-67 which equips the F/A-18s. It is planned to retrofit the whole fleet with this set.

During the late 1980s the whole Mirage fleet received the Tracor AN/ALE-40 flare/chaff dispensers, both launchers being located under the horizontal stabilator on both sides of the fuselage. For air-to-air operations the F1s carry two AIM-9Ns, which are in the process of being supplanted by the improved AIM-9JULI, and in 1996 radar-guided MATRA Super 530F-1s will be received. For air-to-ground operations the Mirages carry dumb bombs on underwing and underfuselage racks; these bombs weigh from 125 to 900 kg (275 to 2,000 lb) and include slick or parachute-retarded, GBU-10 and GBU-16 LGBs, and Mk 20 Rockeye and BME-300 cluster weapons.

Since January 1992 Ala 14 has been responsible for the original Spanish F1 fleet, so all examples have received that wing's fuselage numbers. First- and second-line maintenance is performed at the Escuadrón de Mantenimiento, while modifications and general overhauls are performed at the co-located Maestranza Aérea (Air Rework).

Due to the continuing delays dogging the EF2000 programme, which is the intended F1 replacement, it was decided to begin an upgrade programme for the 55 aircraft (30 CEs, 17 EEs, three BEs, four Cs and one B) remaining. Three French companies made bids: Dassault Aviation, Thomson-CSF and SAGEM. Dassault retired early from the competition and the Logistic Command subsequently declared SAGEM the winner. According to the terms of the contract, this company will supply the upgrade kits while technical personnel from the Maestranza Aérea will undertake the actual work, including a prototype and a pre-series aircraft to be completed in 24 months, with the 53 remaining aircraft being completed by 1999. In order to not adversely affect the unit's operational capability, no more than six aircraft will be in the upgrade line at any time. Alterations to the nav/attack system include new Have Quick II radios, wide-angle HUD with video recording capability, digital IFF/SIF with Mode 4 Crypto, UFC panel, and a DDI in place of the radar display, plus the incorporation of HOTAS controls, digital mission computer, ring laser gyro INS with GPS and weapons management system, the whole to be integrated by a MIL-STD 1553B databus.

141 Escuadrón of Ala 14, which was activated on 18 June 1975, operates F1CEs, EEs, B/BEs or Cs according to the mission requirements. This unit is assigned permanently to Mando Operativo Aéreo. Its main tasking is clear weather interception (CWI), but once the Super 530F-1s are received the unit will be declared AWX. Secondarily, the squadron undertakes FBA (fighter-bomber), AI (air interdiction) and EW (electronic warfare) duties.

142 Escuadrón was commissioned on 1 April 1980, and is tasked with CWI and FBA missions. The unit is also the declared Mirage F1 OTU, training new pilots posted to both squadrons, as well as those of 462 Escuadrón and, until quite recently, those of 111 Escuadrón. To be declared LCR (limited combat ready) a pilot has to perform 88 sorties, 28 simulator missions and many hours of classroom training. Once in the assigned squadron he continues his operational training, and about 18 months later should be declared CR3. Since 1986, 142 Escuadrón, whose badge is a sabre-toothed tiger, has been a full member of the NATO Tiger Association. Both squadrons operate from a complex of 22 hardened aircraft shelters, sharing QRA duties of the Alert Section since 1978, under which two Mirages are kept on 24-hr QRA armed with two Sidewinders and the two 30-mm cannon.

462 Escuadrón, part of Ala Mixta 46 and based at Gando air base, provides fighter cover for the Canary Islands. It was activated on 8 March 1982, and for about 10 years was the owner of all the F1EEs (plus two twin-stick BEs), but this changed from January 1992, when it lost the two-seaters and most of the EEs. Assigned to 462 Escuadrón now are four F1EEs and four CE/Cs which operate from a complex of 10 HASs, one of the aircraft being kept in a QRA shelter.

When 111 Escuadrón of Ala 11, based at Manises near Valencia, phased out its Mirage IIIEE/DEs in October 1992, it received eight Mirage F1CEs loaned from Ala 14. In at least two instances the unit also operated two EEs to qualify for AAR operations. That situation affected the resources of the fighter wing at Los Llanos, so the prospect of purchasing additional aircraft was examined.

This resulted in the acquisition of the Qatar Emiri air force fleet of 11 Mirage F1EDAs and two F1DDAs, which Qatar intended to replace with Mirage 2000-5s. Four single-seaters and a two-seater were ferried from Doha to Manises by 111 Escuadrón pilots on 21-23 August 1994, while two other EDAs were brought to Manises on 1 September 1994 from the Dassault overhaul facility at Bordeaux. The contract included a simulator (received in 1995), Super 530F-1 missiles, Remora ECM pods and spare engines. The remaining six ex-Qatari Mirages will be delivered during the last quarter of 1997. In November 1994, 111 Escuadrón returned four of the loaned F1CEs to Ala 14, thus maintaining a strength of six F1EDAs, one F1DDA and four F1CEs – the latter will be kept until the outstanding ex-Qatari machines are received.

The EF2000, in which Spain has a 13 per cent share, is progressing slowly. The Ejército del Aire has planned to acquire 87 examples of this type, including 16 two-seaters, from 2001 to 2013. It is expected that Spain's FSD aircraft (DA06) will fly towards the end of 1995 or early in 1996.

Maritime patrol and transport

221 Escuadrón of Grupo 22, Ala 21, is based at Morón and is the Ejército del Aire's maritime patrol unit. It was activated in December 1963 as 601 Escuadrón, equipped with the amphibious Grumman SA-16B Albatross at Jerez air base. In 1973 three secondhand P-3A Orions were acquired, one of which was written off in 1977. In 1978 four more were leased from the US Navy, three of which were returned to the US after P-3Bs were received from Norway in 1988/89. In 1992 the unit relocated to the much larger Morón air base, and 221 currently has on strength five ex-RNAF P-3B 'Super Bravo' Orions and two P-3As. One of the Orions is detached on rotation to Sigonella, Sicily, to participate in Operation Sharp Guard. Recently the P-3Bs have received AGM-88 Harpoon missiles for ASW operations. IRA overhauls are made at OGMA facilities in Portugal.

The Ejército del Aire has seven transport squadrons in three wings. The real workhorse of these units is the Lockheed C-130 Hercules, a total of 12 of which were delivered from 1973 to the then-301 Escuadrón at Zaragoza, which later grew into Ala 31 and from August 1989 became Grupo 31 which, with Grupo 15, forms Ala 31. Hercules deliveries totalled seven C-130Hs and five KC-130H transport/tankers. In May 1980 one of the transports was written off in a fatal crash on the island of Gran Canaria, and was replaced in 1987 by a C-130H-30.

Since the FLA is still very much in the future, the Air Staff has developed some plans to increase the capability of the transport units which include acquiring within the next few years between two and six additional Hercules, and upgrading the existing fleet, including fitting a 'glass cockpit' with new avionics, ECM self-protection systems and structural reinforcements. General overhauls are made at the OGMA works in Portugal, where the typical lizard camouflage is being replaced by overall medium grey. The 12 Hercules are shared by Escuadrones 311 and 312, the first specialising in cargo and the latter in tanker/cargo operations. Each crew flies some 450 hours per year, of which 60 per cent comprise tactical and logistic airlift, 15 per cent tactical air

Above: Latest equipment with Ala 48 is a pair of photo-survey Cessna 560 Citation Vs, flown by Esc 403.

Below: Esc 403 is also responsible for Ala 48's U.9s. Both Dornier- and CASA-built aircraft are designated U.9.

Escuadrón 803 is Ala 48's dedicated SAR unit. Its SA 332B Super Pumas (HD.21) are currently being repainted in this overall grey scheme. Previously, they were painted white, but with the same high-visibility yellow bands. Esc 803 shares a primary SAR tasking with Esc 802, based at Gando, in the Canary Islands.

The SAR-configured TR.12A Aviocars of Ala 48's Esc 403 are superficially similar to the specialist EW CASA C.212s of Esc 408. The SAR aircraft, however, lack the fintip radome of the former and carry an AN/APS-128 sea search radar in their 'duck nose'. Nine of these aircraft, which carry the factory designation CASA C.212S1, are in service with Esc 801 and 803 (Ala 11).

Grupo 74's second element is Aviojet-equipped Escuadrón 744. This squadron undertakes refresher jet training for air force staff officers.

Esc 421, of Grupo 42, operates Beech B-55 Barons (E.20). Five of the seven originally delivered are still in use. Ala 42's Piper Navajos and Aztecs have recently been retired.

Grupo 74, based at Matacán, is a training unit with one squadron of CASA Aviojets and one of Aviocars. The 10 T.12B Aviocars of Escuadrón 745 serve a multi-engined trainers for future CN.235 and C-130 pilots.

The second type on strength with Grupo 42 is the Beech F.33 Bonanza (E.24A), of which 18 are in use with Escuadrón 422. Until 1981, this unit was equipped with T-6 Texans.

transport, 15 per cent air-to-air refuelling and 10 per cent special missions on behalf of the UN. The 'Dumbos' – which is their very appropriate radio callsign – have been around the world undertaking humanitarian missions, including Sarajevo, Spilt and Zagreb in the former Yugoslavia. Two KC-130Hs were deployed to Aviano late in November 1994 to support the F/A-18 detachment in Operación Icaro.

Utility transports

Ala 35, based at Getafe to the south of Madrid, began to re-equip with the CASA CN.235 in 1988, and deliveries were completed in 1994. The unit's two squadrons, 351 and 352, share a total of 20 CN.235-100Ms, two of which are in VIP configuration. The remaining 18, which are employed as tactical transports, were delivered in lizard scheme and are being repainted in the omnipresent grey. The CN.235s replaced a similar number of C.212 Aviocars, which in turn were transferred to Ala 37's two flying squadrons (371 and 372) at Villanubla air base, near Valladolid, thus phasing out their old Caribous. The CN.235, which is nicknamed 'mini-Hercules', is very well liked by its crews for, besides featuring modern avionics and navigation systems, it has a good loading capability for its size and is very manoeuvrable. The missions undertaken are the same as those made by Ala 31's Hercules, with the exception of inflight refuelling, and they have ranged far and wide in support of the UN. One of the aircraft is configured as an ambulance for evacuating wounded troops of the Spanish UNPROFOR contingent in Bosnia.

Ala 37 received the 20 CASA C.212 Aviocars, including four in VIP configuration, from Ala 35. In total the Ejército del Aire received 79 Aviocars, 53 of which were of the transport variant. The two squadrons of Ala 37 perform the daily tasking of a transport unit. One of their aircraft is detached to Vicenza air base in Italy, HQ 5th ATAF, for light transport and liaison duties on behalf of Operation Deny Flight. One of these Aviocars was hit by a Serb militia-launched SA-7 in March 1994; some of the passengers were wounded, but the aircraft was repaired quickly and resumed its duties.

Canary Islands

461 Escuadrón, part of Ala Mixta 46 and based at Gando air base in the Canary Islands, is equipped with 11 Aviocars, of which one is in VIP configuration. This is the oldest continually serving squadron in Spanish military aviation, having been activated as 1ª Escuadrilla del Desierto in the early 1920s. Its missions include shuttle flights among the several islands which have military garrisons, as well as the usual tactical missions, and secondary support of 802 Escuadrón's SAR missions.

Currently there are three dedicated Ejército del Aire search and rescue squadrons, which operate Puma and Super Puma helicopters plus Aviocar and Fokker F27MPA fixed-wing aircraft. These provide SAR cover for the assigned FIR zones in mainland Spain, the Mediterranean and the Atlantic Ocean, and they have humanitarian SAR as well as combat SAR taskings. SAR aircraft and helicopters are painted in a high-visibility scheme of white with yellow diagonal bands and international orange undersides. Due to the growing importance of combat SAR and special operations

missions, some are receiving a grey scheme although they keep the yellow band.

1 Escuadrón is based at Son San Juan air base, in the island of Majorca, and is equipped with five SA 330 Pumas (a total of seven was acquired and until very recently served with 402 Escuadrón), four SAR-configured C.212-200 Aviocars nicknamed 'ducknose' due to the radome of its APS-128 search radar, and one C.212-200, without radar, in cargo configuration. 802 Escuadrón, which is based at Gando in the Canary Islands, has three Fokker F27MPAs for long-range operations and two Aérospatiale AS 332B Super Pumas. Since the delivery in the early 1980s of 12 Super Pumas, two have been lost, one from this squadron and the other from 801. 803 Escuadrón at Cuatro Vientos has six Super Pumas, three radar-equipped Aviocars and one Aviocar in cargo configuration.

402 Escuadrón at Cuatro Vientos is currently equipped with two Super Pumas and four Cougars, and is tasked with VIP and government transportation roles as well as SAR and combat SAR missions. This unit, together with 803 Escuadrón and 403 Escuadrón, form Ala 48. 403 Escuadrón is the Ejército del Aire's dedicated photo mapping and aerial survey squadron, being equipped with six Aviocars fitted with De Wilde cameras and two recently delivered Cessna 560 Citation Vs for the same roles.

Air force firefighters

Grupo 43 is based at Torrejón but during the summer months is deployed to many airfields around the country. It is probably the most widely-known Ejército del Aire unit, its yellow and red Canadair CL-215s and CL-215Ts relentlessly fighting forest fires. This unit is unique in that it performs operational missions continually, especially during the summer season. Although it is crewed by Ejército del Aire personnel the aircraft were acquired with funds from the Ministry of Agriculture and Fishing. The unit was activated at Getafe in January 1973 as 404 Escuadrón, relocating to Torrejón several weeks later. In May 1980 the squadron renumbered as Grupo 43.

Since then a total of 30 of the unique Canadair water bombers has been acquired, seven of which have been lost performing this dangerous job. Fifteen have been converted to CL-215T configuration, which entails replacing the P&W R2800 CA3 reciprocating engines with PW123F turboprops, (which provide an additional 1,000 shp/745 kW), along with other refinements such as winglets. Two CL-215s were sold to the Italian company SISAM and, of the other six, one has been donated to the Museo del Aire (Air Museum) at Cuatro Vientos and the remaining five will continue to serve, although if there are enough funds they will be upgraded to the Turbo standard. If not, they will be phased out in about two years' time. During the high season, from June until October, Grupo 43 provides aircraft detachments to seven secondary airfields, covering the mainland, and the Balearic and Canary Islands.

Grupo 45 was formed during the early 1970s as 401 Escuadrón and is equipped with five Falcon 20s for VIP missions and calibration of air bases and airport navigation aids. The unit was based at Barajas-Madrid International Airport until recently, when the unit moved to Torrejón after the base was vacated by USAFE. Current strength consists of three Falcon 20s, two equipped with advanced

electronic systems to check navigation aids and one in VIP configuration; one VIP Falcon 50; two VIP Falcon 900s; and three Boeing 707s, one in VVIP configuration and the other two in tanker-transport configuration equipped with wingtip Sargent-Fletcher refuelling pods. Grupo 45 has two flying squadrons, 451 Escuadrón which operates the Boeing 707s and Falcon 900s and 452 Escuadrón which is in charge of the Falcon 20 and 50 operations.

Electronic warfare systems

In addition to the three Falcon 20s currently on strength the unit also had another two, which in 1994 were transferred to 408 Escuadrón. This squadron, which is the Ejército del Aire's dedicated electronic warfare unit, was activated at Getafe as 408 Escuadrilla (Flight) during the mid-1980s. Its initial equipment consisted of two specially modified Aviocars which were used for training fighter pilots and controllers in ECM avoidance tactics, but also were used as Elint/Sigint/Comint platforms. After receiving two specially modified Falcon 20s from Grupo 45, the unit was upgraded to squadron status and moved to Torrejón and, due to the highly classified nature of its work, established a security fence around its hangar and operations building. Undoubtedly, the star of 408 Escuadrón will be the modified Boeing 707 which has been refitted in Israel with advanced equipment for Elint/Sigint/Comint. This aircraft will have not only intelligence-gathering/reconnaissance capability, but also incorporated into the expensive 'Santiago' programme is a stand-off jamming capability and C³I functions. Ground reception units are being established at the MoD situation room, as well as at new underground hardened WOCs (War Operations Centres) being built at Los Llanos and Morón. Programme manager has been Israeli Aircraft Industries, with Elta providing the electronic equipment in collaboration with the Spanish companies Ceselsa and Indra. The system is expected to attain IOC by late 1996.

Ejército del Aire's Centro Logístico de Armamento y Experimentación (CLAEX – Weapons and Test Centre), based at Torrejón, is tasked with testing new aircraft and weapons for the air force, as well as developing new software for the Hornet fleet. CLAEX is modifying all the AIM-9J/N/P missiles in stock to the AIM-9JULI 'all-aspect' variant which has the AIM-9L seeker head, canard winglets and engine. In the near future its major task will be testing the EF2000 and releasing it for first-line service. Some of its pilots have undertaken test pilot courses in England, France and the United States. The centre's Grupo de Ensayos en Veulo (Flight Test Flying Group) has a small fleet comprising two T-35 Tamíz, two Aviocars and three Aviojets; also on semi-permanent loan are an instrumented F/A-18A+ from Ala 12 and an F1EE from Ala 14.

The Ejército del Aire's Academia General del Aire is located at San Javier air base, in the province of Murcia, south-east Spain. The academy was officially activated on 28 July 1943. The academy year starts in September and ends in July, and about 65 to 70 per cent of the pupils will attain their pilot's wings.

The academy offers two options to would-be air force pilots, both men and women. The first is the Superior Career course lasting five years, with the first two years concentrating on academic

Ala 31, at Zaragoza, is the air force's second 'combined wing'. Its teeth are the two F/A-18A+ squadrons of Grupo 15, Esc 151 (call sign 'Ebro') and Esc 152 (call sign 'Marte'). The wing badge is a turning, twisting tiger with the motto 'quien ose paga', ('who dares will pay for it').

Grupo 15 is currently involved in Operation Deny Flight. Prior to this deployment, some of its aircraft and crews were dispatched to Red Flag to hone their SEAD skills. Ala 12's Esc 124, the Hornet operational transition unit, will soon transfer to Grupo 15, becoming Esc 153.

Grupo 31's Hercules are trading in their unique camouflage for an overall grey finish, as each aircraft undergoes an MLU. This also includes the addition of new avionics, self-defence systems and structural reworking.

Grupo 31 provides the transport muscle for Ala 31. Its Hercules operate in support of the Zaragoza Hornets in the same way the Grupo 45 Boeing 707s do, at Torrejón. Since autumn 1992 the C-130Hs and C-130H-30s (T.10 and TL.10) have been sustaining an airbridge from Torrejón to Split, in support of Spanish UNPROFOR troops.

Escuadrón 312 undertakes the Hercules' air-to-air refuelling mission, with five KC-130Hs (TK.10). The first of these was delivered in 1976, replacing Boeing KC-97s.

Grupo 22 has five ex-Norwegian air force P-3B 'Super Bravo' Orions in service at its Morón home, some 50 km south of Seville. The sub-hunters fly alongside the Aviojets of Grupo 21, as part of Ala 12, but the CASAs are slated for replacement by second hand Hornets.

Two Lockheed P-3A Orions remain in service with Grupo 22, which operates as part of Ala 21, at Morón. These older 'Deltic' Orions are chiefly used for crew training.

Ala 11 operates a mix of Mirage F1s from its base at Manises. The wing has one component squadron, Esc 111. The five ex-Qatari F1EDAs are designated C.14C.

Six single-seat Mirage F1EDAs and a lone two-seat F1DDA (C.14C and CE.14C respectively) are in service with Esc 111. These former-Qatari air force fighters retain their desert camouflage and will be joined by the six currently surviving Qatari F1s in 1997.

matters, the third incorporating the Elementary Flying Syllabus with the T-35 Tamíz trainer, and the fourth the Basic Flying Syllabus with the C.101 Aviojet. (Those pupils unable to attain the necessary flying qualifications at this stage transfer to navigation training.) The fifth year is spent with the Specialisation Schools: Ala 23 for fast jets, Ala 74 for multi-engine conversion and Ala 78 for helicopters. Each year, a handful of the best pupils are sent to America to join the fast-jet course operated by the USAF. Pupils are selected for the fighter/attack, helicopter or multi-engine streams according to their qualifications, personal preference and air force requirements; many of the fast-jet course graduates are posted to the academy as instructors. These specialised courses last for seven to nine months and afterwards the pupils return to San Javier for the graduation ceremony, leaving the academy with the rank of lieutenant.

The other option is to follow the Intermediate Course, from which the student graduates with the rank of sub-lieutenant. He or she can only rise to the rank of Lieutenant Colonel. Students then undergo a three-year course and are able to take flying training in the second year.

The academy is divided into three groups: Grupo de Apoyo for support, Grupo de Material for maintenance and Grupo de Estudios for instruction. This latter group has a Department of Flight and Navigation Techniques which is in charge of the primary, basic and navigation flight schools. The Escuela de Vuelo Elemental has an aircraft inventory of 36 T-35 Tamiz (40 were assembled in Spain by CASA). The pupils undertake a total of 48 flight hours split in two phases – Primary Selective and Primary Advanced – plus 12 classroom hours. In the following year they pass to Escuela de Vuelo Básica which has 42 C.101 Aviojets on strength (the Ejército del Aire received 88 of these jets) performing a total of 111 hours' flying, split into four phases including transition, formation flying and IFR training, plus the fourth stage which comprises 50 classroom hours for studying the aircraft and its systems and 34 simulator hours.

'Patrulla Aguila'

Thirteen of the 42 Aviojets at San Javier are painted in the 'Patrulla Aguila' scheme. Aircraft and pilots for the display team are provided by the Basic Flight School, while the ground crew comes from the Maintenance Squadron. Its strength comprises 14 instructors and seven as reserves. Although 'Patrulla Aguila' is Spain's national air display team it is not a full-time role for the people involved; as is the case with other European teams, instructors and aircraft must perform their primary role within the Basic Flight School. Rehearsals take place two days a week, during the afternoon. The 'Aguilas' (Eagles) perform three different routines depending upon weather conditions and visibility – the flat display has a 1,000-ft (305-m) ceiling, the low display has a 2,500-ft (760-m) ceiling and the high display has a 4,500-ft (1370-m) ceiling. June 1995 marked the team's 10th anniversary, and the Aviojets were painted in their current scheme in 1992.

The Escuela de Navegación has on strength eight Beechcraft F-33A Bonanzas (the air force acquired 30 of these veteran and reliable aircraft), and three Aviocars, two modified as flying classrooms and one as a cargo carrier. Those pupils

who are unable to continue with their pilot training continue their careers at the navigation school and, once qualified as navigators, they work on RF-4C Phantoms, C/KC-130H, P-3A/B Orions, SAR Aviocars, etc. There is a 70-hour two-phase stage in the Bonanza, the first called visual and the second basic, learning IFR navigation techniques. This is followed by 77 hours of advanced navigation in the Aviocar. In addition to the tasks so far described, the Flight Department also has the responsibility of training pilots posted to the academy as instructors.

Fast jet schooling

Ala 23, based at Talavera air base near Badajoz, is the fast-jet training school, being activated on 10 December 1953 as Escuela de Reactores. In March 1953 it received its first six T-33As, and in January 1959 the first F-86F Sabres were delivered. The supersonic F-5B conversion trainers were delivered from November 1970 to an eventual 28 machines, out of a total of 34 F-5Bs acquired by the Ejército del Aire (a total of 70 Freedom Fighters had been ordered, including 36 F/RF-5A single-seaters). In 1991 the remaining four F-5Bs serving with Ala 21 at Morón were also transferred to Talavera.

After graduation from Escuela de Vuelo Básico, future fast-jet pilots can follow one of two routes within Ala 23. The complete course lasts nine months and the flying syllabus comprises a total of 94 sorties plus academic training. The second option is the short course and is undertaken by those pupils who have undergone the fast-jet course, some 110/120 hours with the USAF in the T-38. After returning to Spain, these students spend three months at Talavera, during which they fly 41 sorties.

Ala 23's present strength is 22 F-5Bs and six F/RF-5A (those phased out by Ala 21 in October 1992). It has a secondary tactical CAS role and to act as laser illuminators (with a laser designator in the rear cockpit) while the single-seaters are also tasked for target towing (TDU-10/B target dart) for air-to-air gunnery practice. By the end of 1995, all 22 twin-stick examples will have received an upgrade by CASA, with the kits supplied by Bristol Aerospace Ltd of Canada, including avionics and structural strengthening – making the airframes good for an additional 4,000 hours of operations.

Helicopter units

Ala 78 is in charge of basic/advanced helicopter training for the air force and the army, and until very recently the navy, as well as for Spanish civilian security agencies. It was activated at Cuatro Vientos airfield in 1960, transferring 20 years later to its present location, Armilla air base some 15 km (9 miles) south of the city of Granada with two flying squadrons. 782 Escuadrón is equipped with 15 – of 17 originally acquired – Hughes H-269C (TH-55), and with them future helicopter pilots undergo 50 flight hours and 156 classroom hours. After that they go to 781 Escuadrón which is equipped with eight Sikorsky S-76Cs, the last of which was received in May 1993, the unit having been previously equipped with UH-1Hs. The students there fly 45 hours and undergo 28 one-hour classroom sessions. 781 has a secondary SAR/combat SAR role, constituting 25 per cent of the flight hours assigned to the unit. Also, Ala 78 has a school for training helicopter groundcrew.

Grupo 74, based at Matacán near Salamanca, has one squadron. 745 Escuadrón is currently equipped with 10 Aviocars for multi-engine training for future pilots, who perform some 60 sorties before graduating as transport pilots. This unit was founded during the late 1940s as Escuela de Polimotores. 744 Escuadrón is equipped with 16 C.101 Aviojets and provides refresher training to fast-jet pilots on staff duties. Other tasks include acting as 'targets' during exercises and calibration of military radar sites.

721 Escuadrón, based at Alcantarilla air base, forms part of the Parachute School and is equipped with nine Aviocars. This school provides parachute training to Spanish armed forces and also has a parachute demonstration team. There is a squadron of special forces troops trained as FACs and who, for combat SAR operations, are equipped with light weapons and, recently, equipped with Mistral portable SAMs. Several teams are with UNPROFOR in Bosnia.

The last training unit is Grupo 42, which has two flying squadrons: 421 Escuadrón is equipped with five Beechcraft B-55 Barons (out of seven acquired), and 422 Escuadrón has 18 F-33A Bonanzas. These are used as refresher trainers for transport pilots on staff jobs and for communications and liaison. In the period 1959/61, CASA built 50 Dornier Do 27 liaison aircraft as C-127s, being complemented from 1973 by 25 ex-German Heer and Luftwaffe machines. Some 30 remain, being used at most air bases as liaison and hacks.

211 Escuadrón from Grupo 21, Ala 21 at Morón air base 50 km (30 miles) south-east of Seville, disposed of its 24 F/RF-5As in October 1992. The little Freedom Fighters had ceased to be viable combat aircraft and were transferred to Ala 23 where some half a dozen have been kept in operational status, mainly as target tugs, while the remainder, most with their airframe lives expired, have been broken up as spares. As an interim measure 211 Escuadrón re-equipped with 18 C.101 Aviojets, 10 coming from Escuela de Vuelo Básica at San Javier, and the remaining eight from 744 Escuadrón at Matacán. Although 211 is a fighter unit its task is known as tactical training, and by mid-1996 the unit will start to re-equip with ex-US Navy F/A-18A Hornets.

Ejército del Aire Units
(By Regional Commands)

UNIT	TYPE	DESIGNATION	BASE
Mando Aereo del Centro (MACEN)			
Ala 12			
Grupo 12			
121 Escuadrón	F/A-18A+/B+	C.15/CE.15	Torrejón
122 Escuadrón	F/A-18A+/B+	C.15/CE.15	Torrejón
123 Escuadrón	RF-4C	CR.12	Torrejón
124 Escuadrón	F/A-18A+/B+	C.15/CE.15	Torrejón
	Do 27/C-127	U.9	
Grupo 43	CL-215/T	UD.13	Torrejón
Grupo 45			
451 Escuadrón	B-707	T.17	Torrejón
	Falcon 900	T.18	
452 Escuadrón	Falcon 20	T.11	Torrejón
	Falcon 50	T.16	
CLAEX			
Grupo 44	T-35 Tamíz	E.26	Torrejón
	C.212 Aviocar	T.12B	
	C.101 Aviojet	E.25	
	F/A-18A+	C.15	
	Mirage F1EE	C.14B	

Standing guard outside Ala 23's Talavera home is this North American F-86F Sabre (C.5), which flew alongside T-33s with Ala 23, from 1959. The last of the Ejército del Aire's 270 Sabres was retired in 1973.

Basic helicopter training for the air force and army is the task of Armilla-based Grupo 78. The diminutive Hughes H-269C (HE.20) is flown by Esc 782. Fifteen helicopters provide the initial 50-hour rotary-wing training course.

The Sikorsky S-76 is more often thought of as an executive helicopter, but in the hands of Esc 781, Grupo 78, it is used for advanced training. For this role its twin engines and IFR capability are most useful.

Esc 781 takes the students that have qualified from the basic helicopter course and introduces them to its eight very capable S-76Cs (HE.24) and a 48-hour flying course. The sleek Sikorskys finally replaced Bell UH-1Hs in 1993.

After a crash, due to wing structural failure, the Spanish SF-5 fleet was grounded and a major upgrade programme initiated. This entailed a wing overhaul at Bristol Aerospace in Canada and the entire airframes (combined with new avionics) were then reassembled by CASA. Some refurbished SF-5Bs were redelivered in the 'old-style' silver scheme.

Above: Twenty-two SF-5B trainers remain in service with Ala 23. They are now being repainted in this overall grey scheme, more compatible with their secondary combat role.

Below: Six former Ala 21 S/RF-5As are in use with Ala 23 at Talavera. They have had their cameras and cannon removed and are flown mainly by the instructors.

Above and below: As part of the Academia General del Aire, the Escuela de Vuelo Elemental flies the CASA-assembled T-35 Tamíz (E.26), from San Javier. The five-year 'Superior Career' training course entails two years of study, followed by a year's basic flying on the E.26s. If they make it beyond this point, prospective fast jet pilots graduate to the Academy's Aviojets.

408 Escuadrón	C.212 Aviocar	M.12D	Torrejón
	Falcon 20	TM.11	
	B-707	TM.17	

Ala 35

351 Escuadrón	CN.235	T.19B/C	Getafe
352 Escuadrón	CN.235	T.19B/C	Getafe

Ala 37

371 Escuadrón	C.212 Aviocar	T.12B/C	Villanubla
372 Escuadrón	C.212 Aviocar	T.12B	Villanubla

Ala 48

402 Escuadrón	SA 332B Super Puma	HT.21	Cuatro Vientos
	AS 532 Cougar	HT.21A	
403 Escuadrón	C.212 Aviocar	TR.12A	Cuatro Vientos
	Citation V	TR.20	
	Do 27/C-127	U.9	
803 Escuadrón	C.212 Aviocar	D.3A/B	Cuatro Vientos
	SA 332B Super Puma	HD.21	

Grupo 74

744 Escuadrón	C.101 Aviojet	E.25	Matacán
745 Escuadrón	C.212 Aviocar	T.12B	Matacán

Grupo 42

421 Escuadrón	B-55 Baron	E.20	Getafe
422 Escuadrón	F-33A Bonanza	E.24A	Getafe

Mando Aereo de Levante (MALEV)

Ala 31

Grupo 15

151 Escuadrón	F/A-18A+/B+	C.15/CE.15	Zaragoza
152 Escuadrón	F/A-18A+/B+	C.15/CE.15	Zaragoza

Grupo 31

311 Escuadrón	C-130H/H-30	T.10/TL.10	Zaragoza
312 Escuadrón	C/KC-130H	T.10/TK.10	Zaragoza
	Do 27/C-127	U.9	

Ala 11

111 Escuadrón	Mirage F1C/CE	C.14A	Manises
	F1EDA	C.14C	
	F1DDA	CE.14C	
	Do 27/C-127	U.9	
801 Escuadron	SA 330 Puma	HT.19	Son San Juan
	C.212 Aviocar	D.3A/B	

Mando Aereo del Estrecho (MAEST)

Ala 21

Grupo 21

211 Escuadrón	C.101 Aviojet	E.25	Morón

Grupo 22

221 Escuadrón	P-3A/B Orion	P.3	Morón
	Do 27/C-127	U.9	

Ala 23

231 Escuadrón	F-5A, RF-5A, F-5B	A.9/AR.9/AE.9	Talavera
232 Escuadrón	F-5A, RF-5A, F-5B	A.9/AR.9/AE.9	Talavera
	Do 27/C-127	U.9	

Ala 78

781 Escuadrón	Sikorsky S-76C	HE.24	Armilla
782 Escuadrón	Hughes H-269C	HE.20	
	Do 27/C-127	U.9	

Academia General del Aire

Escuela de Vuelo Elemental	T-35 Tamíz	E.26	San Javier
Escuela de Vuelo Básica	C.101 Aviojet	E.25	San Javier
Patrulla Aguila	C.101 Aviojet	E.25	San Javier
Escuela de Navegación	F-33A Bonanza	E.24A	San Javier
	C.212 Aviocar	TE.12B	
	C-127/Do 27	U.9	

Escuela Militar de Paracaidismo

721 Escuadrón	C.212 Aviocar	T.12B	Alcantarilla

Ala 14

141 Escuadrón	Mirage F1CE	C.14A	Los Llanos
	Mirage F1C	C.14A	
	Mirage F1EE	C.14B	
142 Escuadrón	Mirage F1CE	C.14A	Los Llanos
	Mirage F1C	C.14A	
	Mirage F1B/BE	CE.14	
	Do 27/C-127	U.9	

Mando Aereo de Canarias

Ala 46

461 Escuadrón	C.212 Aviocar	T.12B/C	Gando
462 Escuadrón	Mirage F1C/CE/EE	C.14A/C.14B	Gando
802 Escuadrón	SA 332B Super Puma	HD.21	Gando
	Fokker F27MPA	D.2	

Above: Aviocars are in use for pilot and navigator training, but Esc 761 uses its aircraft for a more exhilarating task.

Below: E.25 Aviojets are the Ejército del Aire's primary jet trainer, and serve with the Escuela de Vuelo Básico, at San Javier.

The Air Academy's TE.12B Aviocars are attached to the Escuela de Navegación. They operate alongside Beech Bonanzas and Dornier Do 27s.

Below: Spain's national aerobatic team is the Aviojet-equipped 'Patrulla Aguila', which functions as a separate entity of the Air Academy, at San Javier.

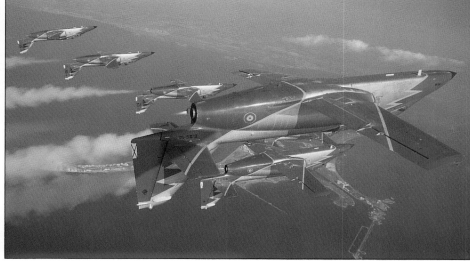

During 1994, four ex-French air force Mirage F1Cs and a single F1B were delivered to Ala 14. They share the C.14A designation of Spain's own Mirage F1CEs.

Below: An Ala 14 Mirage F1CE (C.14A) formates on an F1EE (C.14B). The C.14B carries a Barax ECM pod and ALR-300 radar warning receivers on the tail.

Arma Aérea de la Armada (Spanish Fleet Air Arm)

The Flotilla de Aeronaves (Naval Air Force) is the aerial component of the Spanish Arma Aérea de la Armada (Navy Fleet Air Arm), which is based at Rota with seven escuadrillas (squadrons) Naval Air Station. Together with the carrier *Príncipe de Asturias* and the five Spanish-built FFG frigates, it forms Battle Group Alpha.

The genesis of the current Fleet Air Arm came about from 1917 to 1936 when a Naval Air Service was created and consolidated. Unfortunately, it was disbanded after the Civil War started in July 1936 and its personnel and equipment served on both sides. During the post-war years there were some failed attempts to recreate the service. In November 1954 at Marin heliport in north-west Spain the 1ª Escuadrilla was finally commissioned, equipped initially with three Bell 47G-2 helicopters. Since 1957 the Arma Aérea has been based at Rota NAS, which since then has continually expanded. It boasts three fixed-wing Escuadrillas and four rotary-wing Escuadrillas, with a total strength of 23 aircraft and 36 helicopters, while some 700 men and women serve in the Flotilla. When detached to the Príncipe de Asturias they form an Air Group or Unidad Aérea Embarcada.

Pilot training in the USA

All the future pilots are trained in the United States with the US Navy. After a small 10-hour orientation syllabus with the Hughes 500 of 6ª Escuadrilla they are sent to the United States, where over the course of 22 months they learn the theoretical and practical aspects of flight. Future Harrier pilots perform the complete fighter/strike course, which includes the primary syllabus with 75 hours on the T-34C, basic with 110 hours in the T-2C including carrier qualifications, and advanced in the TA-4J with 110 hours (gunnery, bombing, air combat tactics, etc). Both the Buckeyes and the Skyhawks are in the process of being supplanted by the new T-45 Goshawk. Pilots selected for helicopters log some 110-130 hours in the T-34C in the primary syllabus, followed by 137 hours in the advanced syllabus in the Bell TH-57B/Cs, and if they are posted to 5ª or 10ª Escuadrilla they will get their operational training with the appropriate US Navy replacement air group squadrons.

In addition to the seven Escuadrillas listed in the order of battle, another three existed until recent years but have been disbanded. The pioneer 1ª Escuadrilla, still with the Bell 47G/OH-13s, was deactivated on 30 June 1987. 2ª Escuadrilla was equipped with the first Spanish Navy ASW helicopter, the Sikorsky HRS-2s, the last two of which were withdrawn from use in March 1978 and the unit deactivated. 7ª Escuadrilla, equipped with Bell AH-1G HueyCobras, was deactivated on 19 October 1984, although two of the helicopters continued flying until 1987 attached to 3ª Escuadrilla.

The presently established Fleet Air Arm squadrons are as follows:

Assault 'Hueys'

3ª Escuadrilla – This unit was commissioned on 7 October 1964 and received four Agusta-Bell 204Bs, the first turbine-powered helicopters for the Navy. One of these machines performed the first landing and take-off from the Spanish Navy carrier *Dédalo* (ex-USS *Cabot*) on 19 December 1967. On 31 October 1978, the four 204Bs were phased out of service. From 12 February 1974, 3ª Escuadrilla strength had increased with the arrival of the first Agusta-Bell AB 212 Twin Hueys, a total of 14 machines being received by 1980. Four of them incorporated a datalink system, while another quartet received a Mk 200 electronic warfare set in addition to their ASW equipment. The unit has 10 AB 212s on strength.

Taskings of the AB 212 are varied, the most noteworthy being support for the Marines, logistic transport, SAR, light attack (with two XM200 19-round 70-mm FFAR pods and two MG3 7.62-mm machine-guns), laser illumination with Ferranti 306 designators, electronic warfare and casualty evacuation. The anti-submarine and anti-surface capabilities are not used. The conversion course to the AB 212 is set at 27 sorties with 44 hours to be qualified as co-pilot, while to be rated as aircraft commander requires 250 hours on type or 600 in total. To operate from 'steel decks', either the *Príncipe de Asturias* or any other ship with the capability, the pilot must requalify in landings and take-offs, both day and night every six months. This is a standardised procedure for any Escuadrilla with these capabilities.

First and second echelon inspections are carried out at Rota, while the deep overhauls are performed by AISA (Aeronautica Industrial) in Madrid. 3ª Escuadrilla maintains a high in-commission rate, with up to 90 per cent of its helicopters operational. The first machine to get a limited upgrade was 01-318, receiving some new equipment in the form of self-defence countermeasures, as well as a new light grey scheme that will gradually supplant the typical sea blue scheme worn by these machines. The AB 212s have accompanied the Amphibious Group ships which have carried Spanish UNPROFOR forces to Bosnia since the start of the peacekeeping operations. Not until at least the next decade will a replacement be sought, which could be the Sikorsky HH-60H.

Liaison unit

4ª Escuadrilla – Commissioned on 21 October 1964, the unit received four Piper PA-24 and PA-30 lightplanes, which were grounded for several years due to a legal dispute with the Air Force about operating fixed-wing aircraft. These were phased out in 1992, and the squadron now has a unique complement of three Cessna Citation II twin-jets. The missions are varied, including VIP transport, fleet logistic support, medical evacuations, photographic/infra-red reconnaissance (for which a belly pod can be attached), ship's radar calibration, target facilities and the twice-weekly surveillance sorties to cover the Atlantic and Mediterranean approaches to the Gibraltar straits. To become an aircraft commander in the Citation, the pilot needs a minimum of 1,000 flight hours or 350 on type, which increase to 1,500 for VIP transportation tasks.

ASW and AEW Sea Kings

5ª Escuadrilla – A nucleus of this unit was activated in February 1965, transferring in November to the US Navy's NAS Key West to begin the SH-3 transition. The unit returned to Spain seven months later, being officially commissioned on 29 June 1966 when the first SH-3D arrived at Rota. Another two were received in that year, plus three in 1967, two in 1972, four in 1974 and six in 1981, all in the Delta variant, although the last two batches had some improvements including a Canadian Marconi LN66HP search radar. Since early in its carrier this helicopter has been dubbed by its crews the 'sacred cow' due to its size and complexity. The helicopters were operated by 5ª Escuadrilla for 29 years, during which time six were lost in accidents. Three examples were dispatched to CASA between September 1985 and September 1987 to be converted to the AEW variant with the Thorn-EMI Searchwater radar, similar to the Royal Navy's Sea King AEW.Mk 2s. After conversion, they received a low-visibility grey scheme in place of the gloss sea blue applied to their brothers.

On 16 December 1987 an upgrade programme was authorised to bring the remaining nine machines to late SH-3H standard, with even more advanced systems, in order to extend their operational life until 2015. At that time, they theoretically should be replaced by the SH-60F or a later, more advanced variant. The upgrade incorporates a structural component involving airframe, transmission, turboshafts, electrical and hydraulic systems and avionics, with a new state-of-the-art instrument panel; sonar, with the late generation AQS-81V supplanting the older AQS-13B; and a new tactical navigator (NAT-5), designed and built by Spanish company Ceselsa. The programme is being carried out at Rota and by CASA.

The main mission of 5ª Escuadrilla's SH-3Hs is anti-submarine warfare, and the helicopters can function as troop transports with the ASW gear removed. SAR cover is one of the usual taskings for the unit. To be rated as a Sea King co-pilot 50 hours on type are needed, which increase to 150 to be qualified as combat ready, while for stepping to aircraft commander a minimum of 300 hours is needed. Usually the crews deploy once per year to NAS Jacksonville, Florida, to log the prescribed flight simulator hours. For the Spanish Navy, the early warning SH-3(AEW) represented a revolution in controlling the air and the seas around the fleet. It typically flies at 1,000-2,000 ft (305-610 m), and the Searchwater radar can detect fighter-sized aircraft at 150-200 miles (240-320 km) and Exocet-type missiles at half this distance. The crew is composed of two pilots and two radar operators.

Lightweight helos

6ª Escuadrilla – Activated on 11 April 1972 to operate the Hughes (now McDonnell Douglas Helicopters) 500ASWs, the unit received 14 of these agile helicopters, of which 10 remain in service. The transition period requires 40 hours for qualification as co-pilot and 250 as aircraft commander. After the ex-USN FRAM II destroyers were phased out the anti-submarine mission was abandoned, but the others continue: training, ship's radar calibration, liaison, reconnaissance, light transport and FAC with the Ferranti 306 laser designators.

All of Spain's half a dozen Mirage F1BEs (CE.14) are operated by Escuadrón 142, the Mirage Operational Training Unit and a NATO Tiger squadron.

Below: Mando Aéreo de Canarias consists of Ala 46. This combined wing maintains three dedicated fighter/interceptor, transport and search and rescue squadrons.

Above: This Mirage F1EE carries a Barax ECM pod. These aircraft can also carry the Syrel podded Elint system.

Below: The teeth of Ala 46 are the F1s of Escuadrón 462, though all Spanish Mirages now carry the codes of Ala 14.

Left: Battle Group Alpha is built around the Príncipe de Asturias which embarks a small but effective air wing.

The Armada's most potent asset is 9ª Escuadrilla and its AV-8Bs. The unit was commissioned in 1987.

Left: A mix of ASW and AEW Sea Kings combined with the Harrier force is the backbone of the navy's air power.

The older AV-8As of 8ª Escuadrilla have almost 10 years of solid operational experience behind them.

151

First Harrier squadron

8ª Escuadrilla – This unit was commissioned on 29 September 1976, creating a solid operational doctrine for Harrier operations from carrier decks, which was beneficial to other users of this unique STOVL warplane – the Spanish Navy was the first to operate them in a regular manner from carrier decks. Seven AV-8As and two TAV-8As are on strength, and the squadron's main mission is fleet air defence, tactical strike and reconnaissance, for which there is a wide selection of weaponry available. For self-protection the AV-8As have the Marconi Skyguardian 2000 RHWS. The basic and advanced operational conversion to the Harrier comprises about 100 flight hours. Each pilot has to be carrier qualified and qualified in air-to-air refuelling operations (from Spanish air force KC-130Hs from Grupo 31 and KC-707s from Grupo 45). The aircraft receive first and second echelon maintenance at Rota and third echelon at CASA's factory in San Pablo, near Seville. The name 'Matador' was officially adopted for Spain's AV-8s, but is little used by pilots or groundcrew.

The end of the AV-8A

In 1995 Spain's AV-8As celebrated 25,000 flying hours, chalked up over their 19-year career. Special titles were painted on the fins of most of the aircraft to commemorate the occasion.

The nine AV-8As and TAV-8As will be phased out when CASA delivers eight new AV-8B Harrier Plus aircraft, and will be sold to the Royal Thai navy. Thailand's decision to establish a naval air wing to support the Thai Marine Corps has been an 'on-again-off-again' one, but Spanish shipbuilder Empresa Nacional Bazán is now building a carrier for Thailand which will be similar to, although slightly smaller than, the *Princípe de Asturias*. This vessel, named HTMS *Chakkrinareubet*, is under construction at Bazán's El Ferrol shipyard.

The future of 8ª Escuadrilla is not at all certain, but there is a possibility that it could be deactivated and its personnel transferred to 9ª Escuadrilla, or that the AV-8Bs from the latter would be transferred to 8ª to allow it to continue as an operational unit while 9ª converts to the McDonnell Douglas AV-8B Harrier II Plus.

Second-generation AV-8s

9ª Escuadrilla – The unit was commissioned on 29 September 1987. On 6 October that year the first three of a total of 12 AV-8Bs arrived at Rota after a nine-hour five-minute non-stop ferry flight from St Louis, refuelling eight times from USAF KC-10 tankers. 9ª Escuadrilla has on strength 10 aircraft. To become combat ready a pilot newly posted to the squadron has to make 30 simulator hours and 100 hours on the AV-8B. Before that syllabus was implemented, new pilots underwent a basic stage with USMC VMAT-203 at Cherry Point (some 10-12 hours), but during the last two years the squadron has taken its new pilots from 8ª Escuadrilla. This makes the process easier and less expensive, as the requested TAV-8B two-seat trainer was finally cancelled.

The missions assigned to the squadron are similar to those of 8ª, with the advantage that the Bravo version is much more advanced than the first-generation Alpha. Sixty to seventy per cent of its time is dedicated to fleet air defence, carrying a maximum of four AIM-9M Sidewinders plus the remaining weapons used by the A-8A and the highly effective GAU-12 25-mm gun. For self-protection the Bravo has the ALR-67 radar warning receiver, ALE-40 flare/chaff dispensers and the ALQ-167 active jammer, fitted in a pod on the centreline station.

According to established plans, from early 1996 9ª Escuadrilla will start to receive the first of the eight new AV-8B Harrier II Plus aircraft assembled by CASA. In addition to the APG-65 radar, the Bravo Plus version has a FLIR, more powerful engine, and improved weapons capability including AIM-120 AMRAAM, AGM-84 Harpoon, AGM-88 HARM and AGM-65G Maverick. There is a strong possibility that the 10 straight AV-8Bs will be upgraded to Plus standards so that both squadrons will have a common model, although if 8ª is deactivated all aircraft will remain with 9ª. Since 25 April 1987 an advanced AV-8B simulator built by Ceselsa has been operational at Rota where, besides the prescribed 30 hours for new pilots, operational pilots practice their year's assigned hours simulating emergencies, gunnery, carrier landings/launches in IFR, etc. The AV-8B Plus simulator is scheduled for delivery in 1996.

Spanish Seahawks

10ª Escuadrilla – The unit was commissioned on 5 December 1988 at the same time that the six Sikorsky SH-60B Seahawks arrived at Rota aboard the *Princípe de Asturias*. These advanced helicopters participated aboard 'Santa María'-class (FFG) frigates in the naval blockade of Iraq in the Gulf War, recording 600 flight hours in the Red Sea and the Persian Gulf during those 11 months. A total of 13 pilots and seven sensor operators served in the detachment aboard the frigates *Santa María* and *Reina Sofía*. The Seahawks and their parent frigates continue to participate in Operation Sharp Guard in the Adriatic.

New pilots posted to the unit perform a six-month basic and tactical syllabus with HSL-40 'Air Wolves' at NAS Mayport, Florida, including simulator training. Pilots already operational make yearly refresher visits to this simulator, a situation that will end in mid-1995 when Ceselsa delivers a last-generation SH-60B simulator to the Fleet Air Arm at Rota. After returning from the United States, new pilots perform 50 additional flight hours to be declared combat ready and occupy the co-pilot seat as Air Tactical Officers. The rest of the crew constitutes the pilot (Aircraft Commander) and the Sensor Operator. The squadron's missions comprise anti-submarine and anti-surface warfare, SAR, casualty evacuation and logistics transport, although the Seahawk's primary missions are the two former, forming a fully integrated weapons systems known as LAMPS III with the FFG frigates. According to the original schedule, which is being followed, the unit will receive six more SH-60Bs. First and second echelon inspections are carried out in Rota, while third echelon will be done by CASA.

Carrier backbone

The *Princípe de Asturias* (R-11) was based on a US Navy design for the Sea Control Ship (SCS), a multi-role and relatively low-cost ship. The US Navy abandoned the project in favour of the bigger 'Nimitz'-class nuclear-powered supercarriers but the Spanish Navy became interested in the design, which it considered to be an ideal way to replace the SNS *Dédalo* (ex-USS *Cabot*). The project was acquired and EN Bazán shipbuilder was in charge of developing and building it. Several changes were made, including the 12° ski-jump plus additional accommodation for flag staff, as *Princípe de Asturias* would function also as a command and control ship with Battle Group Alpha and as the Spanish Navy's flagship.

The ship was laid down on 8 October 1979, launched on 22 May 1982 and commissioned on 30 May 1988. Fully loaded it displaces 16,700 tons and is 642 ft (196 m) long, while the flight deck's length is 574 ft (175 m) and has two elevators for aircraft and another two for weapons. The maximum aircraft/helicopter capability is 29 machines, 17 in the hangar deck and 12 on the flight deck. For self-defence it has four Meroka 20-mm CIWS (Close-in Weapons Systems) mounts, as well as chaff launchers and electronic countermeasures systems, while the radar and CIC facilities are very extensive, including digital command and control systems with Link 11 and 14 and secure communications. Main power is provided by two General Electric LM2500 gas turbines which develop a maximum of 46,000 shp (34315 kW) using a single shaft with a six-bladed propeller, maximum speed is 26 kt (48 km/h) and unrefuelled range is 7,500 nm (13875 km) at 20 kt (37 km/h). The complement is 500 men and women, including the Battle Group Alpha flag staff, plus another 250 for the Air Group.

NAS Rota

The Battle Group is based at Rota Naval Station – forming the same base complex as Rota NAS – which is also the home of Fleet HQ. The group's composition varies, with a minimum of four frigates (of the FFG and/or modified 'Knox' classes), the carrier and a oceanic multi-purpose supply ship forming it. The Battle Group is at sea some 120 days per year on average. The composition of the Air Group (Unidad Aérea Embarcada – UNAEMB) changes according to operational needs, and it has its own staff, including GCI controllers, which are installed in the ship's GCI. Each carrier-deployable Escuadrilla has a pilot qualified as LSO who, contrary to conventional carriers, has his work station within Primary Control. A typical UNAEMB composition could be five AB 212s, six SH-3Hs, two SH-3H(AEW), four AV-8As and four AV-8Bs. As mentioned, this changes as requirements dictate.

The six FFG frigates are *Santa María* (F-81), *Victoria* (F-82), *Numancia* (F-83), *Reina Sofía* (F-84), *Navarra* (F-85) and *Canarias* (F-86), and all have hangar decks capable of holding two SH-60Bs. Besides Rota Naval Air Station, naval stations at Marin in north-west Spain (home of the Naval Academy) and Cartagena in south-east Spain have support facilities for helicopter detachments.

Arma Aérea de la Armada units

UNIT	EQUIPMENT	DESIGNATION	BASE
3ª Escuadrilla	AB 212 Twin Huey	HA.18	NAS Rota
4ª Escuadrilla	Citation II	U.20	NAS Rota
5ª Escuadrilla	SH-3D/H Sea King	HS.9	NAS Rota
	SH-3AEW Sea King	HS.9	
6ª Escuadrilla	Hughes 500ASW	HS.13	NAS Rota
8ª Escuadrilla	AV-8A/TAV-8A	VA.1/VAE.1	NAS Rota
9ª Escuadrilla	AV-8B Harrier II	VA.2	NAS Rota
10ª Escuadrilla	SH-60B Seahawk	HS.23	NAS Rota

Above: This 8ª Escuadrilla TAV-8A (VAE.1) was named 'the shark' due to its record of 'eating' FOD.

Below: In 1987 three SH-3s were converted for AEW duties and fitted with Thorn-EMI Searchwater radars.

Below: 3ª Escuadrilla flies the AB.212 Twin Huey (HA.18), chiefly on transport and SAR duties. It is hoped that Sikorsky HH-60Hs will be acquired as replacements in the near future.

Above: All of Spain's 13 remaining SH-3D/Gs have been upgraded to SH-3H standard. They can carry AS.15TT anti-ship missiles, and operate under the callsign 'Morsa' (Walrus).

Below: Four of 3ª Escuadrilla's AB.212s have received an EW/ESM fit, but are rarely used in this role. The fleet is slowly being repainted in a 'low-viz' grey scheme.

Below: In 1988, 10ª Escuadrilla accepted its initial batch of six SH-60B Seahawks, which arrived at Rota on board the Principe de Asturias. A further six SH-60s are on order. They operate from the carrier and Spain's four (soon to be five) FFG-7 frigates. More Sikorskys, in the shape of SH-60F Ocean Hawks, may be purchased to succeed the Sea Kings.

Fuerzas Aeromóviles del Ejército de Tierra FAMET – (Spanish army aviation)

Army flying in Spain developed slowly during the post-war period and it was not until 1958 that the first aircraft specifically acquired for Army support arrived. These comprised two Sikorsky H-19s, two Hiller H-23s and 12 Cessna O-1 Bird Dogs. The aircraft were operated by the Ejército del Aire's 99 Escuadrilla on behalf of the Army, which possessed no qualified pilots at the time. In 1959 several officers were sent to the US Army Aviation Training Center at Fort Rucker for training and two years later seven fixed-wing and four rotary-wing pilots were detached to 99 Escuadrilla to gain operational experience.

FAMET forces are headquartered at Colmenar Viejo base, with 2,277 personnel and 170 helicopters. It forms part of the Army's Rapid Reaction Forces and has nine flying units in six bases and a permanent detachment. Prospective FAMET pilots (both officers and NCOs) pass an initial screening stage at the Colmenar Centro de Enseñanza de las FAMET – CEFAMET (FAMET Training Centre) with medical, physical and technical tests. Those who pass this phase go to the Ejército del Aire Helicopter School, Ala 78, at Armilla, Granada where, after 50 flight hours in the Hughes TH-55, they receive a helicopter pilot rating. After this has been accomplished, the pilots return to CEFAMET for a further 50 hours, this time in the OH-58A, BO 105 and UH-1H, mainly in tactical flying, as well as IFR, finally getting their rating as Army helicopter pilots. Then they are qualified as second pilots and posted to an operational flying unit, where after roughly two years they could be rated as a full Army Combat Pilot.

Today, first-line units are designated as Batallones (Battalions) – one attack, four field (Maniobra) and one heavy. There are three support units at Colmenar, headquarters, training, maintenance and services.

FAMET lineage

On 10 July 1965, the Aviación Ligera del Ejército de Tierra – ALET (Army Light Aviation Group) was established at Colmenar Viejo near Madrid, to establish a force of Army helicopter units and develop systems and tactics. Between September 1966 and January 1967, 14 helicopters (six UH-1Bs, six OH-13Gs and two Agusta-Bell AB 47B-1s) were received by ALET's first unit, Unidad de Helicópteros (UHEL) XI.

A second unit was formed at Colmenar early in 1971 although it did not make its first Huey flight at Rota NAS until 2 August, as the aircraft had arrived by sea from the US only weeks earlier. This unit was charged with supporting the army in the troubled Sahara territory. In December 1971, the unit, by now designated UHEL II, embarked on the Spanish Navy carrier *Dédalo* for the journey to the Spanish Sahara. On board were 15 new UH-1H Hueys which were later joined at the unit's El Aaiun base by three OH-58A Kiowas. In 1975 three SA 319B Alouette IIIs, armed with SS-11 wire-guided missiles or rocket pods and a 20-mm cannon, were also taken on charge. UHEL II stayed in the Spanish Sahara for four years until its return to Spain in December

1975 following the political settlement in which Spain handed over its North African territory to Morocco and Mauritania. During its troubled stay in the desert UHEL II performed several combat missions, even having an SA-7 fired against a Huey, fortunately without hitting it. After the agreement Spanish troops were evacuated from several garrisons, and UHEL II was withdrawn from the territory on 18 December 1975. Its 20 helicopters – an Alouette III had been damaged in an accident and had been previously ferried out – flew in two formations to Lanzarote in the Canary Islands, where they landed on the dock and later were loaded aboard the amphibious ship *Galicia* for the journey to the mainland.

FAMET arrives

While UHEL II was in the Sahara, planned expansion of ALET continued and on 20 March 1973 the service was redesignated Fuerzas Aeromóviles del Ejército de Tierra – FAMET (Army Airmobile Forces). Two more helicopter units, UHEL I (which replaced UHEL XI) and UHEL V, were formed at Colmenar. This heralded a period of rapid expansion which continued until the late 1980s and included the receipt of 60 UH-1Hs, 19 CH-47C/BV-414s, five OH-13Hs, 12 OH-58As, five AB 206As, six AB 212s, 72 BO 105s and 18 SA 332 Super Pumas, in addition to the existing six UH-1Bs, six OH-13Ss and three Alouette IIIs.

In 1979 the MBB BO 105 was chosen over the Improved OH-58, Gazelle and A.109. The AH-1S had been ruled out because of its cost and its inability to double as a liaison and light transport machine. The deal called for the assembly of 60 helicopters by CASA at Getafe, from knock-down kits supplied by MBB. The 60 machines were supplied in three variants: 28 BO 105ATHs (anti-tank), 18 BO 105GSHs (ground support), and 14 BO 105LOH (light observation). The BO 105ATH is armed with six wire-guided HOT missiles; the BO 105GSH has a belly-mounted Rheinmetall RH202 20-mm cannon with a 550-round magazine in the aft cabin. The BO 105LOH carries no armament. Prior to deliveries of CASA-assembled machines from September 1980, FAMET acquired 11 ex-German Army BO 105s from the evaluation batch. After refurbishment, these arrived at Colmenar late in 1979 to begin the training of air and ground crews for the new type. An MBB BO 105P demonstrator was also loaned and later acquired, making a total of 72 of these helicopters acquired by FAMET.

Attack unit

Most of the BO 105s, including all the ATHs, were delivered to BHELA-1, which was created at Colmenar in June 1980. Batallón de Helicópteros de Ataque I (Attack Helicopter Battalion I) is FAMET's only attack unit and is deployed at Almagro base, in the province of Ciudad Real. By 1983 the unit had reached full strength in personnel and machines, by then having 45 BO 105s and one UH-1B, and relocated to the newly constructed base of Almagro, some 200 km (125 miles) south of Madrid. Current

strength is 28 ATHs, a number of LOHs and five GSHs. The battalion's flying organisation comprises three anti-tank companies (1ª Cia C/C 'Halcón', 2ª Cia C/C 'Lagarto', and 3ª Cia C/C 'Lince'), each equipped with six ATHs and one LOH. The remaining BO 105s are kept in reserve and in maintenance rotation, with the exception of six LOHs used by the Headquarters Company and four GSHs used by an Armed Reconnaissance Section, plus another example on reserve.

According to Spanish Army modernisation plans, and if budgets permit, BHELA I is scheduled to begin re-equipping with 35 ex-US Army AH-64A Apaches from mid-1998. The BO 105 will receive a limited upgrade and continue in service in reconnaissance, liaison and light attack duties.

After arriving from CEFAMET, a prospective BHELA I pilot must fly 85 hours to be qualified as LCR (Limited Combat Ready). Later he will continue with his operational training which includes simulator sessions and the firing of a HOT missile. Once declared Combat Ready, each pilots fires two missiles per year, and to reach the maximum level of Aircraft Commander (CR3) he needs some 500 hours, or roughly two years.

Following UHEL II's return from the Spanish Sahara, the unit was based at El Copero, Seville, but its helicopters were regularly detached to the Spanish sovereign cities of Ceuta and Melilla on the North African coast. In August 1979 UHEL II moved to a newly constructed base at Bétera, about 20 km (12 miles) north of Valencia. Following a directive some years later, its name changed to Batallón de Helicópteros de Maniobra II – BHELMA II.

Multi-role Huey

The ubiquitous Huey has been the mainstay of this unit and of the FAMET, although its importance will diminish over the next years. Apart from its main roles of cargo and personnel transportation, the UH-1H has been employed for parachuting, medical evacuation and as gunships. FAMET's Hueys use several armament options, including three types of machine-guns, two types of rocket launchers and a 40-mm grenade launcher. Currently, BHELMA II has 15 UH-1Hs which equip two light transport companies, each with six Hueys, while the other three are in maintenance rotation or attached to the headquarters company, as are three BO 105LOH. Six BO 15GSHs form a reconnaissance company. From May/July 1991, six of its UH-1Hs, with two from BHELMA III and two Chinooks, detached to northern Iraq for the UN-sponsored Operation Provide Comfort.

BHELMA II, which celebrates its 25th anniversary in 1996, is in the process of changing its UH-1Hs for the much more advanced Eurocopter AS 532 Cougar. A total of 18 machines are earmarked for the unit, in which three transport companies will be equipped, as the reconnaissance company will transfer its BO 105s to Almagro before the end of 1995 and convert to transport. One of the already established transport

The Cessna Citations of 4ª Escuadrilla can be fitted with this Vinten camera pod for photomapping and infra red reconnaissance duties.

Two Cessna 550 Citation IIs were delivered to 4ª Escuadrilla, in 1983, followed by a third example in 1989. They are the only type now operational with this unit, which until recently had the Piper Comanche and Twin Comanche on charge.

Above and below: 4ª Escuadrilla has now retired its last PA-24-260 Comanche (E.30) and PA-30-160 Twin Comanche (E.31). Both types were the subject of a long wrangle with the air force, which revolved around which service should control fixed-wing aircraft operations.

The Hughes 500ASWs (HS.13) of 6ª Escuadrilla no longer serve in the anti-submarine role and have been relegated to the light transport and training duties. Like the AB.212s, however, they can carry Ferranti 306 laser designators for forward air control missions.

Two BHELTRA V Chinooks were deployed to Turkey for Operation Provide Comfort. Nine of the CH-47Cs have been upgraded to CH-47D/BV 414 standard, with uprated engines and fibreglass rotorblades.

BHELTRA V is the FAMET's heavy lift unit and operates its 18 CH-47C/D Chinooks (HT.17) from Colmenar. Some of these aircraft now carry a Bendix weather radar in the nose.

The Aérospatiale AS 332B Super Puma was acquired in preference to the AB.412 Griffon. CASA assembled 12 of the 18 Pumas delivered, while France acquired five Aviocars as part of a reciprocal deal.

companies will cease operations with the Huey during the second half of 1995 to start the process of converting to the French helicopter at Colmenar and El Copero, later going to France for the Cougar course. It is expected that the first three Cougars will reach Bétera during the first quarter of 1996, while the following 15, which possibly will be assembled by CASA, will be delivered in approximately a year and half after the first deliveries. By then the third company will be completing its conversion period and BHELMA II will be again declared operational in the Army's order of battle.

BHELMA III was activated on 23 May 1974 as UHEL III at Colmenar Viejo, moving in November of the same year to Agoncillo base, in the province of Logroño, northern Spain. It was equipped initially with UH-1Hs and in 1976 received three Alouette IIIs for its reconnaissance section, which were transferred to the Ejército del Aire in December 1981. BHELMA III has a strength of 15 UH-1Hs in two transport companies and six BO 105GSHs and three BO 105LOHs in one reconnaissance company. The unit specialises in mountain flying.

BHELMA IV, activated in July 1975 at El Copero base near Seville, is currently equipped with 15 Super Pumas, shared between one HQ and two transport companies. This unit, as well as other FAMET units, is in charge of maintaining a rotational detachment at the North African cities of Ceuta and Melilla. Super Puma deliveries took place during 1988/89, six delivered directly from Aérospatiale and 12 assembled by CASA in

Getafe, Madrid, of which three are permanently based at Colmenar with Servicio de Helicópteros (Helicopter Servicing Unit). During 1993/94, BHELMA IV disbanded its reconnaissance company and its BO 105s were transferred to BHELA I.

The FAMET's newest unit is BHELMA VI, which was activated on 26 April 1986. Currently, its strength comprises eight UH-1Hs and six AB 212s equipping two transport companies. The unit is based at the military camp of Los Rodeos International Airport, in Tenerife, Canary Islands.

FAMET's heavy transport unit is BHELTRA V (Batallón de Helicópteros de Transporte V), which was activated at Colmenar in 1973 as UHEL V and changed to its present designation in 1980. The unit, known as the 'Pegasos', is divided into two heavy transport companies. Initial equipment was two CH-47C Chinooks, a third being lost in an accident shortly after its arrival, fortunately without casualties. Another CH-47C was delivered in 1973, followed by four in 1974, three in 1977 and two in 1982. During 1984 a further order for of six machines was confirmed, these being of the BV-414 variant, which was a hybrid model with several features of the US Army's CH-47Ds. Two were received in June 1986, two in November and two in April 1987.

BHELTRA V missions have included the transport of Phantom fighters from Torrejón to Getafe for overhaul by CASA, the recovery of smaller helicopters, and many civilian support and humanitarian tasks, although the unit's main role is to provide logistic support for the Army,

particularly the Parachute Brigade. During the late 1980s it was announced that Boeing Vertol would upgrade nine CH-47Cs to the Delta variant; BHELTRA V received the first of these in September 1991, followed by four in 1992, while the final four arrived early in 1993. BHELTRA V now operates six BV-414s, nine CH-47Ds and two CH-47Cs. Upgrading of the CH-47Cs (including the BV-414s) is being considered.

FAMET's three other helicopter units are based at Colmenar. Batallón de Transporte (BATRANS) is a liaison unit and operates four UH-1Hs. CEFAMET (Centro de Enseñanza de las FAMET) is tasked with providing the specialised course for future Army fliers. It has 21 helicopters on strength comprising eight OH-58As, five BO 105LOHs and eight UH-1Hs. Finally, SHEL (Servicio de Helicópteros) has a five-helicopter fleet consisting of three Super Pumas, one BO 105LOH and one UH-1H. The unit loans it aircraft to the centralised maintenance facility at Colmenar (Unidad de Mantenimiento y Apoyo), which is in charge of third-level maintenance of the whole fleet, and also loans them to FAMET's HQ (Jefatura FAMET – JEFAMET) for command and control duties and for VIP transportation.

Future plans, besides the acquisition of Cougars and Apaches and the upgrade of the remaining Chinooks, are centred around the construction of an advanced simulation centre at Colmenar with Chinook and Super Puma simulators; continuation of the NVG acquisition programme including adapting the cockpits of the oldest helicopters to make them compatible with NVG operations; and fitment of secure radios (Have Quick) in the whole fleet, as well as defensive countermeasures, chaff/flare launchers, IR jammers, etc. Finally, FAMET bases will receive state-of-the-art control and approach radars. **Salvador Mafé Huertas**

FAMET Units

UNIT	EQUIPMENT	DESIGNATION	BASE
BHELA I	BO 105ATH BO 105LOH BO 105GSH	HA.15 HR.15 HR/A.15	Almagro
BHELMA II	BO 105LOH BO 105GSH UH-1H	HR.15 HR/A.15 HU.10B	Bétera
BHELMA III	BO 105LOH BO 105GSH UH-1H	HR.15 HR/A.15 HU.10B	Agoncillo
BHELMA IV	Super Puma	HT.21	El Copero
BHELMA VI	UH-1H AB-212	HU.10B HU.18	Los Rodeos
BHELTRA V	Chinook	HT.17	Colmenar
BATRANS	UH-1H	HU.10B	Colmenar
CEFAMET	Kiowa BO 105LOH UH-1H	HR.12B HR.15 HU.10B	Colmenar
SHEL	BO 105 Super Puma UH-1H	HR.15 HT.17 HU.10B	Colmenar

Left: Between 1988 and 1989 15 Super Pumas replaced the UH-1s of UHEL IV, which was expanded to battalion strength in April 1988, as BHELMA IV. The Pumas are all equipped with anti-icing gear for high-altitude operations.

The Super Pumas (HT.21) are allocated to BHELMA IV and the 15 aircraft are divided between two transport companies and one HQ company. Three Pumas are also on permanent detachment with SHEL.

Only six AB.212s fly with the FAMET and all are in service with BHELMA VI (at Los Rodeos in the Canaries) alongside eight Bell UH-1Hs. Activated in 1986, BHELMA VI is the FAMET's newest unit.

Above: BHELA I is the FAMET's attack battalion and relies on 28 HOT-armed BO 105ATHs (HA.15).

Right: Spain's unique cannon-armed BO 105GSHs (HR/A.15) fly in the armed reconnaissance role.

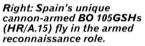

Of the 12 OH-58As delivered, eight remain in use with CEFAMET, the training unit, at Colmenar.

Below: In heavy gunship configuration FAMET's UH-1Hs (HU.10B) can carry two 0.5-in machine-guns.

Above: Between 1971-1978, 60 UH-1Hs were delivered to Spain, equipping virtually all FAMET's flying units. It is hoped to replace them with the Eurocopter Cougar.

Below: Plans to replace the 'Hueys' with 40 UH-60s were abandoned. Instead, the UH-1s have been upgraded with composite blades, new avionics, RWRs and chaff/flare units.

INDEX

Picture acknowledgments

Front cover: Randy Jolly. **4:** Alenia, McDonnell Douglas. **5:** Dassault, Antoine J. Givaudon. **6:** Antoine J. Givaudon, Texas Instruments, Carey Mavor. **7:** Agusta, Aldo Ciarini. **8:** Chris Lofting, Chris Ryan. **9:** British Aerospace, Andrew Marden. **10:** PZL, Jeremy Flack/API. **11:** IAI (two). **12:** Alenia, Carey Mavor. **13:** Patrick Laureau, Westland. **14:** M Slides via Bob Archer, Matthew Leong. **15:** Matthew Olaufsen, Northrop Grumman. **16:** McDonnell Douglas. **17:** McDonnell Douglas, AFMC. **18:** Tim Ripley (two). **19:** Patrick Allen, Westland. **20-22:** Tim Ripley. **23:** Patrick Allen, Tim Ripley (two). **24:** Per Udsen, El-Op, Lockheed Martin. **25:** Joe Cupido, Lockheed Martin. **26:** David Donald (three). **27:** Patrick Laureau (three). **28-39:** Dougie Monk. **40:** Randy Jolly, Mitsubishi. **51:** Mitsubishi, Randy Jolly. **52:** M. Takeda. **53:** Peter R. Foster, Peter Steinemann. **54:** M. Takeda, A.J. Payne. **55:** Peter R. Foster, A.J. Payne (two). **56:** A.J. Payne (two), M. Takeda. **57:** Peter R. Foster, A.J. Payne. **58:** A.J. Payne, Robbie Shaw. **59:** A.J. Payne (two), Randy Jolly. **60:** M. Takeda, A.J. Payne (two). **67:** Robbie Shaw. **68:** US Air Force (two). **69:** A.J. Payne (two), Chris Pocock, Robbie Shaw. **70:** A.J. Payne, Randy Jolly, Robbie Shaw, A.J. Payne, Peter Steinemann. **72-73:** Randy Jolly. **74:** Austin J. Brown/APL, Robert Hewson. **75:** Ted Carlson/Fotodynamics. **76:** Randy Jolly. **77:** John Gourley (four). **78:** David Donald (two), Ted Carlson/Fotodynamics, Bob Archer. **79:** John Gourley (four). **80:** Walter Wright. **81:** John Gourley (seven). **82:** Ted Carlson/Fotodynamics, Randy Jolly. **87:** John Gourley (five). **88:** Walter Wright. **89:** Austin J. Brown/APL, Randy Jolly. **91-93:** Randy Jolly. **94:** Robert Hewson. **95:** Randy Jolly. **96:** Ted Carlson/Fotodynamics, Randy Jolly. **97:** David Donald (three), John Gourley (two). **98:** Randy Jolly (two). **100:** Graham Robson. **101-103:** Randy Jolly. **104:** Norm Taylor via Robert F. Dorr, Robert Hewson, Robert F. Dorr, Bob Archer. **105:** Randy Jolly. **106:** Randy Jolly (three), Robert Hewson, Rockwell. **107:** Ted Carlson/Fotodynamics, Walter Wright, Randy Jolly, Philip Chinnery, David Donald. **108:** Randy Jolly,

Renato E.F. Jones, Bob Archer, Rockwell. **109:** Randy Jolly (three), Walter Wright, John Gourley, Bob Archer, Robert Hewson. **110:** Randy Jolly, Philip Chinnery (two), Bob Archer, Stephen Kill, Alec Moulton, D. Adams. **111:** Robert Hewson (two), Peter R. Foster, Philip Chinnery, Randy Jolly (three), John Gourley (two). **112:** Randy Jolly (four), Robert F. Dorr. **113:** Peter J. Cooper, Rockwell, Randy Jolly (three). **114:** Jeff Rankin-Lowe, Bob Archer, Randy Jolly (two). **115:** Randy Jolly, Henry B. Ham, Ted Carlson/Fotodynamics (two), Jeff Rankin-Lowe (two). **116:** David Donald, Ian Black, Ted Carlson/Fotodynamics, Randy Jolly, Lockheed Martin. **117:** Jan C. Jacobs via Robert L. Lawson, Randy Jolly, Joey A. Zerbe, Stephen Kill. **118:** David Donald (two), Henry B. Ham (two), Hans Nijhuis. **119:** Graham Robson, Michael Grove via Robert L. Lawson, Randy Jolly. **120:** British Aerospace, Steve Hill/EMCS, Henry B. Ham, Michael Grove via Robert L. Lawson. **121:** Randy Jolly (two), Henry B. Ham, via Robert F. Dorr. **122:** James Benson, Henry B. Ham, Randy Jolly, David Donald. **123:** Randy Jolly, David Donald, Alessandro Bon, Gilles Auliard. **124:** David Donald, Ted Carlson/ Fotodynamics (two), Henry B. Ham. **125:** Jeremy Flack/API, David Donald, via Robert L. Lawson, Henry B. Ham, Jody Louviere. **126:** Henry B. Ham (two), Peter R. Foster, Joey A. Zerbe. **127:** Peter Cooper (two), David Donald, Ted Carlson/Fotodynamics, Jeff Rankin-Lowe. **128:** Daniel Soulaine, Jody Louviere, Carey Mavor, M.P. Hopper, Robert Greby. **129:** B. Redfern, Ted Carlson/Fotodynamics, Alec Fushi, Ted Carlson/Fotodynamics, Ted Carlson/Fotodynamics, A. Bakker. **131:** Henry B. Ham (three), Nathan Leong, Ted Carlson/Fotodynamics. **132:** Alec Fushi, Randy Jolly, Chris A. Neill, Matthew Olafsen. **133:** Daniel Soulaine (two), Ted Carlson/Fotodynamics, Alec Fushi, Rick Llinares/Flightline, Regent Dansereau. **134:** Alan Key, Ted Carlson/Fotodynamics (three), Jody Louviere, Randy Jolly. **135:** NJ ANG, Gilles Auliard, Daniel Soulaine, Robert Greby. **136:** Jeremy Flack/API, Jeff Rankin-Lowe, Ted Carlson/Fotodynamics, Daniel Soulaine, Jeff Rankin-Lowe, Robert Greby, Carl Richards. **139:** Salvador Mafé Huertas (four), E.A. Sloot (three), Kevin Wills. **141:** Salvador Mafé Huertas (five), Kevin Wills (three), E.A. Sloot. **143:** Kevin Wills (two), Robin Polderman, Salvador Mafé Huertas (three), E.A. Sloot. **145:** Robin Polderman, Hans Nijhuis (three), Salvador Mafé Huertas (three). **147:** Salvador Mafé Huertas (seven), Stephen Kill. **149:** Salvador Mafé Huertas (seven), Robbie Shaw (two). **151:** Kevin Wills (two), Salvador Mafé Huertas (six). **153:** Hans Nijhuis (four), Salvador Mafé Huertas (six). **155:** Salvador Mafé Huertas (four), Hans Nijhuis (three), René van Woezik. **156:** Yves Debay. **157:** Yves Debay (two), Hans Nijhuis (two), Salvador Mafé Huertas (four).